Investing in Energy

Investing in Energy

Creating a New Investment Strategy to Maximize Your Portfolio's Return

Michael C. Thomsett

palgrave
macmillan

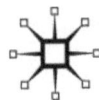

INVESTING IN ENERGY

First published in 2014 by
PALGRAVE MACMILLAN®
in the United States—a division of St. Martin's Press LLC,
175 Fifth Avenue, New York, NY 10010.

Where this book is distributed in the UK, Europe and the rest of the world, this is by Palgrave Macmillan, a division of Macmillan Publishers Limited, registered in England, company number 785998, of Houndmills, Basingstoke, Hampshire RG21 6XS.

Palgrave Macmillan is the global academic imprint of the above companies and has companies and representatives throughout the world.

Palgrave® and Macmillan® are registered trademarks in the United States, the United Kingdom, Europe and other countries.

ISBN: 978–1–137–35716–8

Library of Congress Cataloging-in-Publication Data

Thomsett, Michael C.
Investing in energy : creating a new investment strategy to maximize your portfolio's return / Michael C. Thomsett.
pages cm
Includes bibliographical references and index.
ISBN 978–1–137–35716–8 (alk. paper)
1. Energy industries—Finance. 2. Power resources—Finance. 3. Investments. 4. Portfolio management. I. Title.

HD9502.A2T5176 2014
332.6—dc23 2013026336

A catalogue record of the book is available from the British Library.

Design by Newgen Knowledge Works (P) Ltd., Chennai, India.

First edition: January 2014

10 9 8 7 6 5 4 3 2 1

CONTENTS

Part IV Summarizing the Market

FIGURES

Tables

TABLES

Introduction

ENERGY AS A WHOLE HAS MORE TRADING VOLUME THAN ANY OTHER commodity. Why is this?

You rely on energy for virtually everything you do. You gas up your car, heat and cool your homes and offices, and more. Energy products are used in so many different industries that without energy products, the entire global economy would not be able to function.

This book presents a range of topics that represent the "fundamentals" of investing in the energy market as well as an examination of technical features. The term "fundamentals" refers to the tangible financial and market factors affecting value whereas the technical analysis studies price movement. To invest in any sector, you need both of these.

The book contains four parts. In part I, you get all of the basics you need to get started and to understand the broad range of this market. Part II describes the many ways to trade in energy, including direct ownership of stocks or commodities and "packaged" methods (exchange-traded funds, commodity index funds, and mutual funds, for example). Part III moves beyond the ways to invest and presents specific trading and investing strategies. And part IV summarizes the entire book and offers ideas for focusing your analysis and understanding of this complex and high-volume market.

As you read through this book, remember the primary purpose: To provide you with the tools and range of knowledge you need to make an informed investment decision. Many traders focus on a narrow range of indicators to determine how, where, and when to enter trades. For example, you might focus only on chart patterns or earnings reports to buy or sell stock, on world oil prices to invest in energy commodities, or on seasonal temperatures to time entry into heating and cooling stocks or funds. All of these make sense, but all are only part of the broader picture. This book does not propose any one method over another; its purpose is to present you with a study of the entire market. This study is presented from the point of view of individual investors (or as Wall Street cynically calls them, "retail" investors), those who, like yourself, want to understand the strengths and weaknesses of a particular sector, the range of methods for trading in it, and the risk levels associated with specific products.

Even if you have considerable investing experience and are familiar with the market in general, you will benefit from the information provided in this book. Fundamental analysis tends to focus on individual financial results of companies, and this is a shortcoming of how and why anyone picks products (stocks, bonds, ETFs, commodities) and sectors (energy, for example). However, there is much more. This book focuses on the fundamentals of the sector, which encompass far more than the profit and loss, cash flow, and dividends paid by energy companies. In this regard, the fundamentals are expanded to include the nature of a particular energy product (supply and demand, utilization, geopolitical influences, and other fundamental and economic realities).

Most investors think immediately of the cost and use of oil as the energy market—and for good reason. Oil represents a major share of the worldwide energy market, not to mention the heaviest volume of futures trading. It is the giant of the energy sector. However, people often overlook the many other types of energy products, how they are created, where they are used, and the types of investment opportunities they present.

In the chapters about these individual products, detailed explanations are included about the supply and demand, range of use, and other factors about each topic. The purpose is not only to give background about the product and its range of uses, but to point the way to potential investment opportunities based on evolving supply and demand, new technology, and the scarcity or abundance of alternative energy sources.

This is not a book aimed at suggesting specific trading methods or even a selection of subsectors within the energy market; it is aimed at explaining each of these in enough detail to help you improve your fundamental research and, perhaps, to expand the way you analyze sectors in determining the value of companies and their stocks or commodities.

PART I

OILING UP ON THE BASICS

CHAPTER 1

A Brief History: Black Gold, Texas T

> It takes millions of years to create and seconds to burn—so why do we continue to use oil when it will soon run out?
>
> Duncan Graham-Rowe, "Black Gold," *The Guardian*, April 29, 2008

Is the oil industry a byproduct of the automobile age? Yes and no.

History's record reports examples of asphalt being used to build walls and buildings. Asphalt, a petroleum product, was extracted from sources near Babylon and in Greece. Over many centuries, various oil products have been used, but it wasn't until the nineteenth century in the United States that oil became a major industry.

Kerosene processing and refining was the discovery of a Nova Scotia scientist, Abraham P. Gesner, in 1846. By 1854, petroleum refining was advanced by distillation methods developed at Yale University. However, it was Baku, Azerbaijan, that dominated world oil production by the middle of the nineteenth century:

> By the nineteenth century, Azerbaijan was by far the frontrunner in the world's oil and gas industry. In 1846—more than a decade before the Americans made their famous discovery of oil in Pennsylvania—Azerbaijan drilled its first oil well in Bibi-Heybat. By the beginning of the twentieth century, Azerbaijan was producing more than half of the world's supply of oil.[1]

The first modern-day oil well was developed in 1859 near Titusville, Pennsylvania. The distinction between this and past oil wells is that it was the first drilled well, compared to others that were dug. The well was powered by a steam engine, and it started the first of many American oil booms. Today, oil accounts for 40% of energy use in the United States and for 90% of auto fuel. Perhaps the biggest controversy about the oil industry is found in the question: How much oil is left and when will it run out?

Key point: Use of oil in the United States is higher than just about anywhere else in the world. But this is *not only* due to high levels of consumption. As a leading economic power, the United States also shows high energy use as a symptom of robust economic activity.

Since the beginning of the nineteenth century, this has been the major question, and alarm bells have been going off for the past 100 years, based on predictions that the world is about to run out of oil. As it turns out, estimates of worldwide oil reserves have been revised upward almost as often as predictions of oil's demise have been published.

These reserves are updated not only because new reserves have been found, but also because new extraction methods have made it possible to extract oil from sources once thought inaccessible, such as shale and sand reserves, which are immense.

The operation of OPEC (Organization of Petroleum Exporting Countries) affects oil prices also affects demand for oil, often leading to price increases to create the appearance of shortages regardless of whether these shortages exist or not. Meanwhile, as alternative energy sources are created and as technology continues to improve for all forms of energy, the actual supply and demand issue is made more complex.

PREDICTIONS OF OIL'S FINAL DAYS

Since the beginning of the twentieth century, experts have been predicting that oil is about to run out. It seems reading the future is not all that easy.

Key point: It is easy to make predictions about the future, but accurate predictions are much harder. The only "easy" prediction is about the past.

The following is a partial list of these predictions:

1914: Oil supplies will last only one more decade, according to the U.S. Bureau of Mines.

1920: Only 20 billion barrels of oil exist in the entire world, according to the U.S. Geological Survey (USGS).

1922: Worldwide oil will run out within 20 years (USGS).

1926: The Federal Oil Conservation Board (FOCB) estimated that only 4.5 billion barrels of oil existed in the entire world.

1932: FOCB revised its estimates of worldwide oil supplies to 10 billion barrels.

1944: Only 20 billion barrels of oil remain in the entire world, according to the U.S. Petroleum Administrator for War.

1950: The American Petroleum Institute declared that only 100 billion barrels of oil remained in the earth.

1951: The U.S. Department of the Interior stated that oil supplies would run out by 1964.

1956: Dr. M. King Hubbert predicted a most likely date for an oil production peak was 1965 and that with greater growth the peak year would be 1970.

1962: Dr. M. King Hubbert predicted world oil supplies would peak by 2000.

1972: The UN Conference on the Human Environment reported that "peak production" would occur in 2000. ("Peak" refers to the time when oil production reaches its maximum and then begins to decline.)

1974: Dr. M. King Hubbert revised earlier predictions and estimated peak oil production by 1995.

1976: The U.K. Department of Energy reported that peak production would occur by the end of the century.

1977: President Jimmy Carter predicted that by the end of the 1980s all of the world's oil would be used up.

Dr. M. King Hubbert named 1996 as the year when peak oil production would occur.

1979: Shell Oil Company stated that peak production would occur within 25 years.

Shell Oil's prediction was seconded by the World Bank.

1995: Petro-consultants C. J. Campbell and J. H. Laherrère predicted peak oil production within 10 years.

1996: Published in *World Oil*, expert L. F. Ivanhoe said 2010 would be the peak oil production year.

1998: The International Energy Agency (IEA) stated that peak production would occur by 2010.

1999: The USGS predicted 2010 as the year of peak oil production.

2001: Kenneth S. Deffeyes wrote in *Hubbert's Peak: The Impending World Oil Shortage* (Princeton, NJ: Princeton University Press) that peak production would occur in 2003.

2004: National Geographic reported in a cover story that peak oil could happen as soon as within five years.

2005: John Vidal wrote in *The Guardian* that oil production could peak within a year.

2007: Fredrik Robelius wrote in *Science Daily* that peak production would occur in 2008 under worst-case conditions or in 2018 under best-case circumstances.

2008: The German Energy Watch Group predicted that the peak year had arrived ("Peak Oil is Now," in *EWG Bulletin*, Oct. 30, 2008).

Paul Krugman wrote in the *New York Times* that the peak oil theory had become plausible.

2009: The U.K. Energy Research Centre reported a global peak was inevitable and close.

The U.K. Industry Taskforce on Peak Oil and Energy declared an oil crunch within five years.

The IEA announced that peak oil had occurred.

2010: Kenneth S. Deffeyes wrote in *When Oil Peaked* (New York: Hill & Wang) that peak production had already occurred.

The U.S. Joint Forces Command (USJFCOM) predicted that "By 2012, surplus oil production capacity could entirely disappear, and as early as 2015, the shortfall in output could reach nearly 10 million barrels per day (*Joint Operating Environment 2010*).

Paul Krugman declared in the *New York Times* that peak oil had already occurred.

2012: The IEA stated that its 2009 announcement of peak oil having arrived had been premature.

2013: A huge increase in oil production quieted predictions, for the moment. Ben German wrote in "BP CEO: 'Peak Oil' Talk Quieted by Abundance" (*The Hill*, March 6, 2013) that production, including onshore shale and deepwater reserve tapping, had shown past predictions of production peaks to be wrong.

It seems apparent at first glance that oil will eventually peak and then decline. However, as new reserves, methods, and technology are developed, the predictions of impending peak oil have been shown to be far off. Every limited resource will eventually be depleted, but over the last 100 years an array of "experts" and agencies have demonstrated that their read of oil as a resource has been far off. Why is this?

> **Tip:** Every commodity's demise has been predicted over many decades. Even agricultural shortages and famines have been predicted as recently as 50 years ago. These predictions do not account for innovation and technological advances.

The proven reserves of oil have increased with new discoveries. These have been made possible with advanced technology for locating previously unknown fields, and as prices have risen, even previously proven reserves that were too expensive to extract have been moved into the profitable column. Technology has also made it possible to extract oil from sand and shale reserves, which in North America are vast. And the use of the effective technique of hydraulic fracturing ("fracking") has also created a new boom in oil and gas extraction.[2]

There are numerous reasons for experts and agencies to have misunderstood the supply of gas and oil. The history of predictions reveals, though, that this is an industry that continuously develops new and improved methods for finding and extracting reserves in the earth and even under the oceans.

Hubbert's Peak

One of the best-known theories about the supply of oil is named after M. King Hubbert, a geophysicist who predicted that oil production would peak within a few years, after which the world would begin running out of the resource. The so-called Hubbert's peak was a theory worthy of consideration, acknowledging that limited resources do run out at some point. However, Hubbert made the mistake of naming a specific year when this would occur and then revising his estimate several times. Since the mid-1950s, Hubbert has identified 1965, 1970, 1995, 1996, and 2000 as years when the peak would occur. Contradicting these predictions, actual production of oil has not peaked and started to decline but has steadily increased.

Key point: Making predictions about peak oil can be based on assumptions about what is known at the moment. But if you have to continually revise and update your prediction, your assumptions are called into question.

The U.S. Department of Energy reported in late 2012 that oil output in the United States rose by 3.7% to 6.5 million barrels per day due to improved technology (horizontal drilling and fracking, for example) and that the primary locations for this higher volume were in Texas and North Dakota, where shale oil reserves have led the trend.[3]

Although Hubbert's peak predictions have been proven inaccurate, the theory itself may have merit. The timing of oil's demise, however, appears decades off based on newly evolved technology. The theory itself is based on observations of increased demand versus proven reserves, and a seemingly inevitable point at which demand would exceed the supply level. The reference of Hubbert's peak may be applied regionally, but it usually refers to worldwide oil supplies.

Under the Hubbert peak theory, reserves were limited to oil that could be easily extracted. Advanced mining and extraction methods were ignored in the calculations, but in reality any resource can be extracted with advanced technology, and in recent years, technology has demonstrated that this is so. The theory of a peak also ignores the cost of oil and gas extraction, which is an additional factor in exploiting known reserves. For example, due to environmental restrictions oil located in the Outer Continental Shelf is going to be more expensive to extract than land-based oil fields.[4]

Because reserves vary in cost to acquire, the Hubbert peak theory assumes that the peak refers to easily or cheaply attained reserves and that more expensive sources will not become attractive until after the peak. However, advances in technology bring this assumption into question and are one reason that the estimated peak date has to be continuously pushed back to later years.

The debate raises an interesting question about supply and demand. Precisely how is this economic factor defined in the energy business? If you limit the question to pure economics, supply and demand are easily understood. Prices are determined by the interaction between the two sides; if the resource becomes scarce, prices rise, and if it becomes more plentiful, prices fall. However, this isolated definition of supply and demand has to be expanded to include political and environmental considerations.

Political effects on supply and demand include the geopolitical controls put in place by OPEC. By limiting production, OPEC drives up prices, thus artificially distorting the supply and demand for energy products. On the domestic level, political considerations include a desire by the government to focus on alternative energy and to phase out dependence on oil and gas. In fact, the Obama administration favors ending subsidies for traditional oil and gas exploration and transferring those benefits to the development of alternative energy sources. But is this wise? If development and expansion of energy resources is to rely on proven reserves, that tends to limit potential growth in the future (when additional reserves may be discovered and when technology makes reserves more accessible). Looking back in history, this point was made by a prominent economist:

> Industrial development would have been greatly retarded if sixty or eighty years ago the warning of the [coal] conservationists had been heeded.... The internal combustion engine would never have revolutionized transport if its use had been limited to the then known supplies of oil.... Though it is important that on all these matters the opinion of the experts about the physical facts should be heard, the result in most instances would have been very detrimental if they had had the power to enforce their views on policy.[5]

Tip: If humans were to limit their inventions to resources known at the moment, nothing would get done. Virtually every advance is a combination of imagination and a belief in the ability to improve resources in the future.

In terms of supply and demand, whenever the government attempts to control resources because it favors a different path, the outcomes often are not entirely as intended. The wage and price controls enacted during the Nixon administration make this point. The administration believed it could control inflation with these steps. At the beginning of 1973, inflation was under 4%, but a year later it rose to over 9%. The real consequence of this attempt to control prices was a drastic increase in both inflation and interest rates. Although this approach to the economy has been discredited as a consequence of Nixon-era decisions, the same idea—controlling which type of energy is used by limiting supplies and moving tax incentives—is likely to fail now as well. The economy is resilient, and demand invariably means that oil and gas find a way to the market.

In addition to political actions, a second factor affecting supply and demand is environmental. A great debate has been underway for more than a decade about global warming. Those who doubt the concept do not see how curtailing the use of oil and gas makes sense; those who believe in it see oil and gas as major contributors to pollution.

The environmental considerations—based not only on a belief in global warming but also a desire for clean energy alternatives—translate to federal policy once again. By limiting access to federal lands and the oil reserves they hold, the price of domestic energy is forced upward. This policy has been stated as, "Somehow we have to figure out how to boost the price of gasoline to the levels in Europe."[6]

However, a distinction must be made between a philosophically based policy favoring alternative sources of energy and the economic realities of supply and demand. The desire to raise domestic gasoline prices to levels of Europe (approximately $8 per gallon) is an attribute of the desire to pursue clean energy alternatives, whether they are practical or not. The policy has consequences, as does any attempt to tinker with the economy. An opposing opinion stated that

> hammering the American consumer with high gas prices to make electric and hybrid cars more appealing is consistent with Obama administration policy and Chu's philosophy. That explains the refusal to allow the building of the Keystone XL pipeline and to allow drilling in wide areas of the United States and offshore areas.[7]

The combination of political and environmental influences on the supply and demand factors of energy complicates analysis. It is not a matter of whether people agree or disagree with the policy or whether they accept or reject the notion of global warming. The important issue is that these motivations distort supply and demand domestically just as much as OPEC price manipulation affects energy prices internationally.

Estimating Worldwide Reserves

All good intentions aside, the peak theory gets revised every few years as new reserves are discovered. Proven and known reserves are quite large, and reserves in the United States are among the biggest in the world.

Key point: The reserves of oil and natural gas in the United States are among the largest in the world. However, not all of these reserves are accessible or economically viable for extraction, and not all are allowed to be exploited under current environmental limitations.

Hubbert's peak theory has been accepted by not only one analyst, but it is a political and environmental mantra as well. While the long history of predictions about oil running out demonstrates the problems with predicting the future, the whole question of supply and demand is complicated by the political and environmental assumptions of the government and environmental agencies. Although recent activity in extraction of oil and gas demonstrates that newly discovered reserves bring all past predictions into question (and even disprove them), this is a difficult discussion to have. Keeping the issue to the facts about supply and demand is no easy task, because the debate encompasses opinion, belief, and even wishful thinking—on both sides.

OPEC's Role

The Organization of Petroleum Exporting Countries (OPEC) was founded in 1960 and consisted then of the five founding nations: Iran, Iraq, Kuwait, Saudi Arabia, and Venezuela. Since its founding, other nations have joined as well: Qatar (1961), Libra (1962), the UAE (1967), Algeria (1969), Nigeria (1971), Ecuador (1973), and Angola (2007). One-time members Gabon and Indonesia have suspended their memberships, leaving a total of 12 current members.

The stated mission of OPEC is "to coordinate and unify the petroleum policies of its member countries and ensure the stabilization of oil markets in order to secure an efficient, economic, and regular supply of petroleum to consumers, a steady income to producers, and a fair return on capital for those investing in the petroleum industry."[8] Originally, the stated purpose of OPEC was to ensure the "right of all countries to exercise sovereignty over their natural resources." However, in the 1970s, OPEC began setting policies on price levels and production of oil among member countries in response to the 1973 oil embargo.[9]

Collectively, OPEC member nations produce 81% of the world's oil annually or 1,200 billion barrels, and non-OPEC nations produce the remaining 19% or 282 billion barrels. Each day, these nations produce millions of barrels of oil. Figure 1.1 summarizes OPEC's daily production of oil.

In comparison, non-OPEC countries also produce a large portion of the world's oil. Figure 1.2 shows the daily oil production among the eight highest-capacity non-OPEC countries.

When daily production is compared to levels of proven reserves, you can see the disparity between production and consumption. Figure 1.3 summarizes proven reserves of oil in OPEC countries and the eight largest reserves in non-OPEC countries.

To study consumption worldwide, check figure 1.4, which summarizes the countries consuming over 2 billion barrels of oil per year.

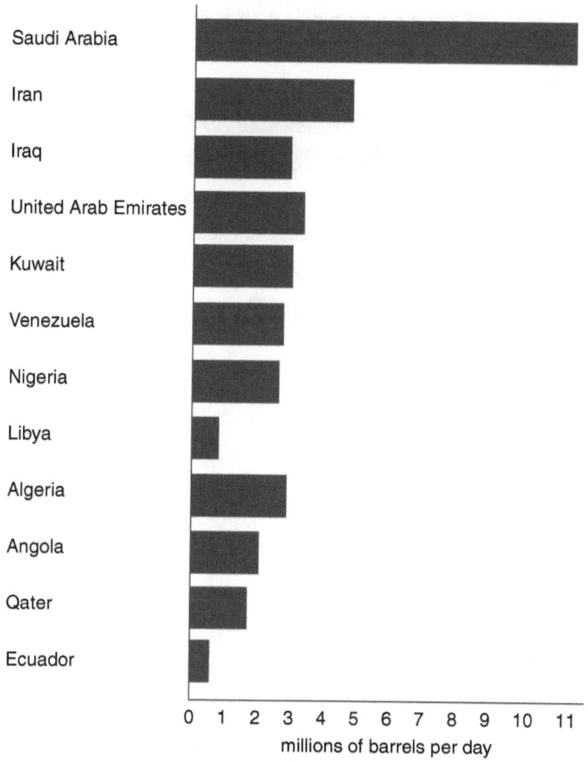

Figure 1.1 Production, barrels per day, OPEC countries
Source: Prepared by author from raw date, OPEC Annual Statistical Bulletin 2012 and U.S. Energy Information Administration (EIA.gov)

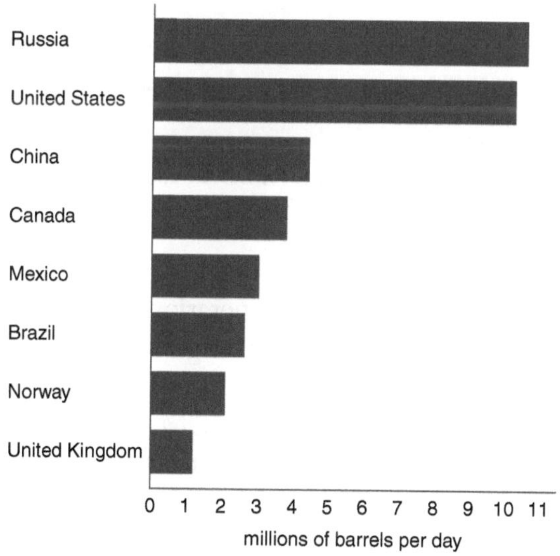

Figure 1.2 Production, barrels per day, 8 highest-volume non-OPEC countries
Source: Prepared by author from raw date, OPEC Annual Statistical Bulletin 2012 and U.S. Energy Information Administration (EIA.gov)

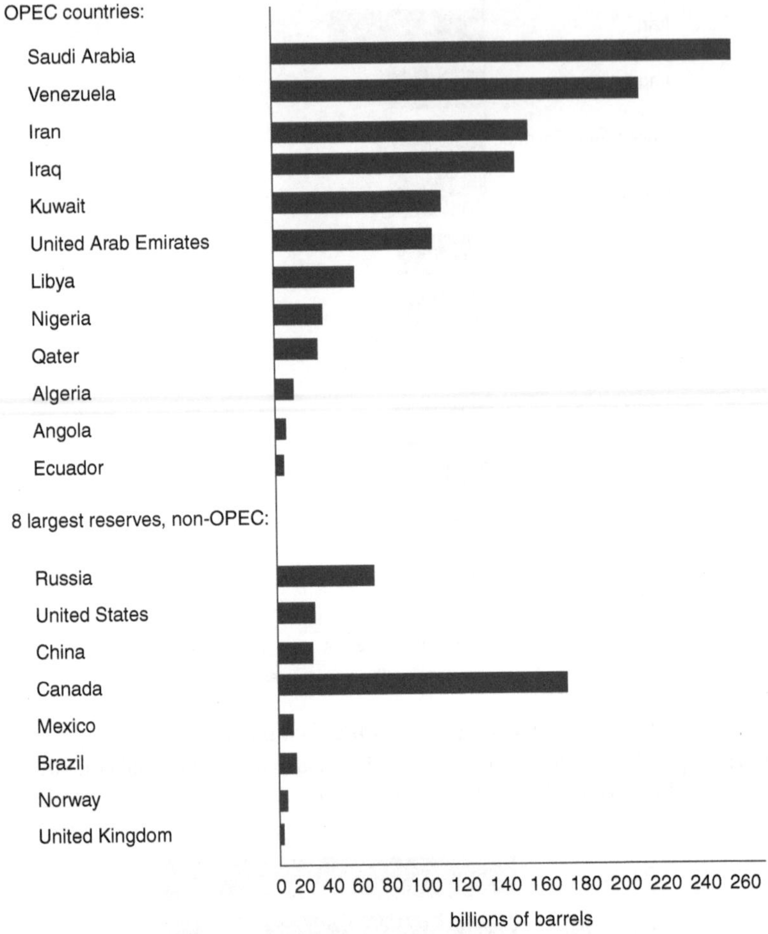

Figure 1.3 Proven reserves
Source: Prepared by author from raw date, OPEC Annual Statistical Bulletin 2012 and U.S. Energy Information Administration (EIA.gov)

Energy usage is a factor in the generation of goods and services, and the United States remains the strongest economy in the world, which is reflected in its oil consumption among many other factors. This brings up another economic fact: Many people believe that the United States is dependent on the Persian Gulf oil producers for most of its oil. But this is far from true. In fact, the United States gets only 13% of its oil from the Persian Gulf, and 74% of total supplies are from the United States, Mexico, and Canada. The sources of oil for the United States are shown in figure 1.5.

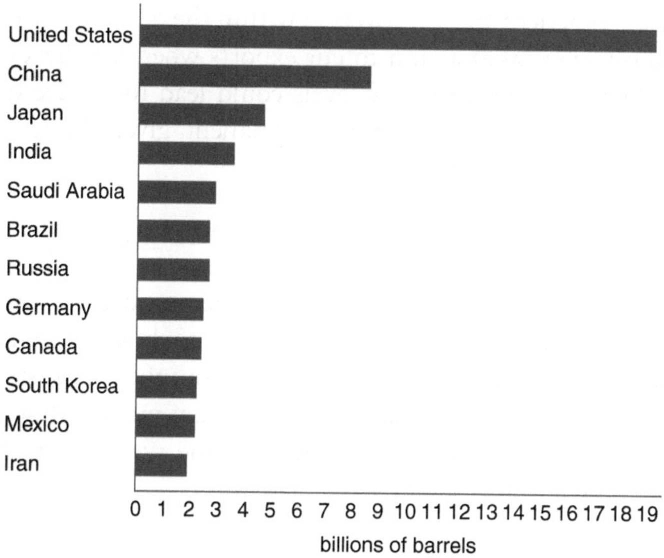

Figure 1.4 Annual oil consumption (countries over 2 billion barrels per year)
Source: Prepared by author from raw date, U.S. Energy Information Administration (EIA.gov)

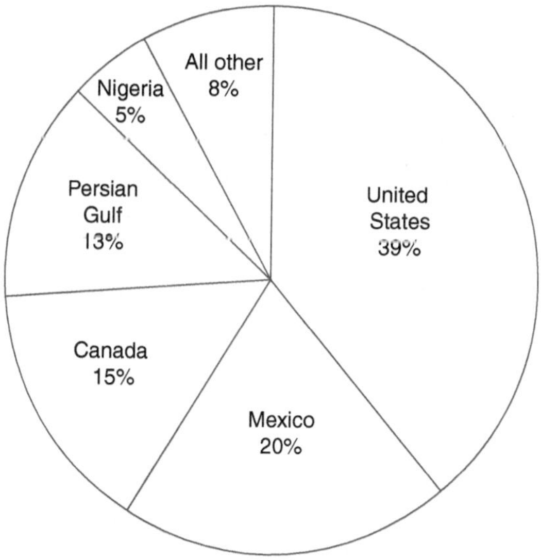

Figure 1.5 United States: Oil consumption by source
Source: Prepared by author from raw date, U.S. Energy Information Administration (www.eia.gov)

OPEC's practices are controversial within the oil industry because the organization takes action to cut exports when it believes prices are too low or when inventory levels could lead to a price decline. However, this trend is by no means permanent, given changes in the global oil market, where

> oil supply is rising outside of OPEC. In September [2012], US oil production hit 6.5 million barrels a day, the highest level in almost 15 years. The high oil prices of the past decade spurred the opening of unconventional resources such as the Bakken Shale, which has boosted North Dakota's output almost sixfold in the past five years. That state alone now pumps almost as much oil as OPEC-member Qatar. To the north, Canada is developing its vast oil sands.[10]

Some predictions have been made that OPEC will not be able to sustain its high price per barrel in coming years due to increased production in the United States and Canada as well as in other non-OPEC nations. With OPEC oil prices above $110 per barrel in 2012, the outlook for the future points to likely lowered price levels. This is due to increased oil supplies in non-OPEC countries:

> Since 2010, prices have remained far above the cost of competing supplies, such as producing oil from shale ($50–75), tar sands ($55–85), natural gas ($70), or coal ($90) let alone deepwater fields. The financial incentive to develop new supplies outside OPEC is overwhelming.
>
> It seems unlikely that OPEC will be able to defend the current price level without accepting a substantial reduction in its market share.[11]

The oil industry is evolving, and while OPEC once controlled world oil prices due to its domination over production, the situation is changing rapidly. For investors and traders, this means that a robust non-OPEC market with competitive prices and new sources of energy products (shale, sand, natural gas, and coal, for example) will change the entire supply and demand picture within the industry.

Tip: Fundamentals of the energy market are as much political as they are economic. As technology advances, today's seemingly fixed supply and demand picture will evolve.

Energy and Politics

The effects of the geopolitical climate on markets for energy cannot be overlooked. More than in any other commodity, energy is subject to many forces outside the simple channels of supply and demand. This applies not only globally but locally as well.

As a fundamental part of the energy market, federal policies have a direct and immediate effect on the supply of and demand for energy products. The Obama administration has stated its policy to encourage alternative sources, such as solar and wind power, as well as demanding improved CAFE (Corporate Average Fuel Economy) standards over the next decade. In August, 2012, the administration set a standard requiring 54.5 mpg by the year 2025. This idea was not without controversy. According to the National Automobile Dealers Association, this standard will translate to $3,000 added to the average price of vehicles. According to the NADA, "This increase shuts almost 7 million people out of the new car market entirely and prevents many millions more from being able to afford new vehicles that meet their needs."[12]

Tip: When standards for performance are set, the outcome is not limited to improved air quality. Related economic consequences should be considered as well.

The energy market is affected by political considerations even beyond questions of supply and demand. A paradox is seen in the increased domestic production of oil that is occurring at the same time that these standards are being tightened. At the same time, oil extraction technology is improving and production has been rising to record levels in the United States and Canada, and OPEC's share of world oil production is on the decline.

Environmental interests have a direct impact on government policies concerning oil production, use, and even levels of extraction and refining of oil. The supply of world oil might peak one day, but it seems that this will not occur any time soon. In spite of Hubbert's predictions and the predictions of other experts, today's supply and demand outlook indicates that the oil supply is here to stay for many decades to come.

Key point: Unlike so many markets involving the study of fundamentals, in the energy market, perception has as much influence—and sometimes more–as the more apparent economics of supply and demand.

In most forms of investment, it is easy to understand how supply and demand work. A producer of an agricultural product has a specific market level that can be tracked over many years and may be accurately predicted into the immediate future. Prices will change based on weather, production, and other factors, but these are variables to a consistent supply and demand market. This means the fundamentals of the market are orderly and logical. But in the energy industry, the fundamentals involve not only the true demand for oil, but manipulation of production and prices by OPEC, curtailment of domestic production by a government wanting to encourage alternative energy sources, and a variety of conflicting environmental and political forces, and all of them influence the picture of supply and demand.

CHAPTER 2

The Big Picture: Energy, the Biggest Market Sector

> Fossil fuels supply 86% of the world's energy—the energy that makes the difference between a life expectancy of 40 years in undeveloped countries and 80 years in industrialized countries. By contrast, a meager 2% of the world's energy is produced by "green" sources such as wind, solar, and biofuels—after over three decades of subsidies around the globe. There is no evidence whatsoever that these sources can provide the combination of attributes necessary to provide industrial-scale power: abundance, high energy concentration, and reliability.
>
> Alex Epstein, "Three Myths about Oil,"
> *Forbes,* June 18, 2010

The range of the energy industry is vast. The best-known form of energy, oil, contributes significantly to most forms of industry, perhaps more than many investors realize. Among the fundamentals of energy investing is awareness of the widespread applications of this resource.

The dependence of the world on energy is undeniable, and this is by no means a new phenomenon. Mention of energy in ancient times includes the use of asphalt for waterproofing and caulking by the Sumerians; it was also used by the ancient Greeks for mortar in the walls of Babylon. In about 3000 BC asphalt was used for waterproofing in India. Even the ancient Egyptians used asphalt in the process of embalming and preserving mummies.[1]

The belief that use of oil and byproducts like asphalt is new is contradicted by these historical examples. However, in modern times, it was during the nineteenth century that the industry grew most in Europe and the United States. Kerosene became a major source for light and heat until it was replaced by electricity, but rather than ending dependence on fossil fuels, the generation of electricity required further use of other energy products to meet production demands. And at the beginning of the twentieth century, advances in automotive manufacturing placed even more demand on the energy market: gasoline to run automobiles, asphalt to pave roads, electricity to power production plants and machinery.

A Diverse Market

In the eight major sectors of the economy, direct application of energy products is essential, for example:

Basic materials and industrial goods: oil and gas are essential to produce most products; even agricultural products rely on oil and gas for manufacture and operation.

Conglomerates: diversified communication and aerospace companies rely on energy products in manufacturing plants, operations, transportation, and fueling machinery.

Consumer goods: appliances, automobiles, rubber, plastics, and paper all depend on energy products for production and operation.

Health care: drug manufacturing, medical equipment, laboratories, hospitals, clinics, and research facilities all require energy products.

Services: freight services, auto dealerships, and computer and electronics stores use energy products for their operations.

Technology: energy products are required for computers, telephones, and semiconductors to operate as well as for the production of components.

Utilities: electricity, gas, and water utilities all rely on massive uses of energy for generation, transport, and operations at every level.

Crude Oil Is 33.5% of World Energy Use

One-third of all energy use worldwide is derived from crude oil.[2] The division of world energy use is summarized in figure 2.1.

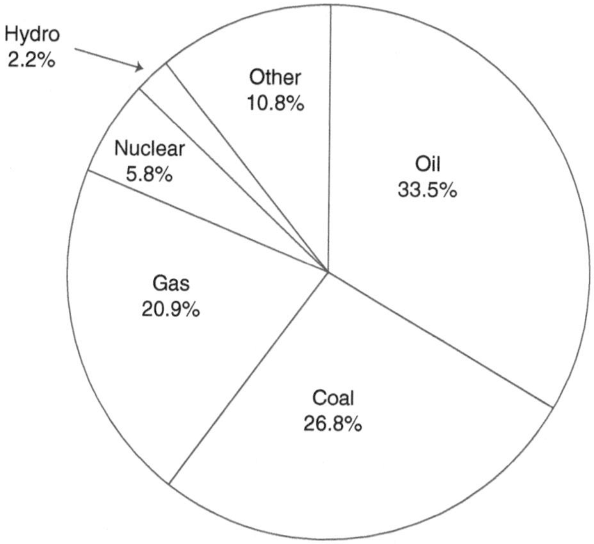

Figure 2.1 Total world energy consumption by source, 1990–2009
Source: Prepared by author from raw data in "Energy in Sweden 2010, Facts and figures," Table 46 "Total world energy supply, 1990–2009

This reliance on crude oil is complicated by the fact that between 1990 and 2008 energy use per person grew by 10%. China experienced a 146% increase, while in the United States energy consumption increased by 20%. This growth concerns primarily crude oil and coal consumption. China has become one of the world's larger consumers of energy (18% of all energy used in 2009), of which 33% was oil and 27% coal.[3]

China also is the world leader in the expansion of renewable energy resources. About one-fifth of global investment in renewable energy was made in China (spending $52 billion); the United States ranked second (spending $51 billion).[4]

> **Tip:** For future investment value, pay attention to where investment is being made today for development of new resources and technology.

The dominance of China in increased energy consumption of all types provides investors with an important fundamental fact for analysis. Companies in China developing energy resources may outperform US-based companies, and these may hold stock in specific companies or shares of exchange-traded funds (ETFs) or commodity index funds. (See chapter 5, "Trading in Energy Futures," chapter 6, "Energy Stocks," and chapter 7, "ETFs and Index Funds.")

In the United States, one reason for lower expansion in crude oil was a widespread restriction in the period of 2008–2013 on oil and natural gas drilling even in areas with abundant reserves. Between 1982 and 2008, one-half of the waters in the Gulf of Mexico were restricted for oil exploration, and most of the Outer Continental Shelf was similarly restricted. Strong opposition to development of reserves in Alaska further curtailed the production of domestic reserves. On September 30, 2008, Congress allowed many of these restrictions to expire, leading to the extraction of at least an additional 18 billion barrels of crude oil per year. However, the requirement of approval of new drilling permit approvals by the Department of Interior has kept these resources out of reach, and the Arctic National Wildlife Refuge's (ANWR) estimated 5.7 to 16 billion barrels also remain restricted (this is the portion within the federal area of ANWR, estimates as of 1998).[5] ANWR may represent 10% or more of total nonproven oil reserves in the United States, which the Department of Energy estimates at 120 billion barrels.[6]

AMERICAN ENERGY POTENTIAL

Contrary to the idea that the United States is running out of oil, the nation is rich in petroleum potential, with untapped reserves in the form of both proven and unproven reserves; some estimates place the oil potentially available in the United States on a par with the reserves in the Middle East.

Many sources of domestic oil have been too expensive for production, and others have been made inaccessible by environmental interests and the federal government. Others still can be accessed with new technologies and remain abundant, awaiting only government approval or private investment to become additional sources of new domestic energy. As oil prices rise and scarcity (real or perceived) troubles the market, many sources once too expensive will become feasible for drilling and extraction.

Key point: Supply and demand are not unchanging factors. Changes in price levels and technology both affect how much oil can be extracted profitably.

A great debate is underway among the Hubbert's peak proponents, those who believe that peak oil production is likely to occur soon or has even occurred already. These groups, including environmental groups as well as scientists, want dependence on oil to go away, but that isn't going to happen overnight. However, are these beliefs truly based on what is best for ensuring clean air and a healthy planet, or are they based on being an anti-oil and anti-business? Both elements are present, and it is not always easy to tell them apart. However, from the point of view of investors and traders, *scarcity* (whether real or created) provides positive benefits in the market. Supply and demand rule all markets, and scarcity is one of the most important fundamental indicators in the energy sector.

How do supply and demand really work? Consider the real estate market. In strong markets supply diminishes, and as a result prices rise. In weak markets demand is low, and inventories of available properties begin to lose value. The same is true in all markets, including energy. It does not matter whether scarcity is real or only perceived; what does matter is how prices react to that scarcity. For example, when political conflict heats up in the Middle East and Iran threatens to block the Strait of Hormuz, the effective outcome is that oil prices rise in the United States. The belief in coming shortages itself creates a perceived supply shortage. The Strait of Hormuz is strategically crucial to world oil supplies. Only 21 miles wide at its narrowest point, it is the only sea lane for the transport of oil from the Middle East. In 2011, an average of 14 tankers passed through the strait each day, representing 35% of seaborne oil and 20% of worldwide supplies.[7]

In spite of the potential of the United States and its untapped energy reserves, even a slight reliance on oil from the Middle East underscores that the goal of energy independence makes sense. The desire for energy independence is not only a matter of control but also of market price and management over scarcity. Investors in other sectors are well aware of how scarcity affects prices. In the agricultural market, a freeze in Florida or a drought in the Midwest have direct and immediate price consequences on agricultural commodities, for example. In 2011, a freeze in Florida was reported and had immediate price consequences: "Orange juice futures hit a three-year high Tuesday after a 4.6 percent jump."[8] In the Midwest, 2013 commodity prices were threatened as well:

> The 2012 drought spread beyond the Midwest to affect more than 60% of the contiguous United States, making it the worst since the Dust Bowl in the 1930s. A sharp drop in crop yields pushed corn and soybean prices to record highs during the summer and costs to feed US livestock soared, forcing ranchers to send their herds to slaughter rather than pay the higher feed costs.[9]

The Good Thing about Price Volatility

In any investment market (including commodities) volatility, uncertainty about the future, makes trading interesting and risky—the greater the risk, the greater the possibility of profits. Oil futures, for example, are the most actively traded and among the most unstable commodities in the world. This attracts speculation.

> **Key point:** Heavy volume of trading in energy commodities tells you that oil is essential not only for the obvious reasons, but also in its secondary uses. In this respect, volatility in trading is a positive attribute.

Energy futures have the highest trading volume of all commodities. The ranking as of the beginning of 2013 is as follows:[10]

1. Crude oil
2. Coffee
3. Cotton
4. Wheat
5. Corn

6. Sugar
7. Silver
8. Copper
9. Gold
10. Natural Gas

As the commodity traded at the worldwide highest volume, crude oil is a highly "liquid" market. This means that oil is easily bought and sold on the market without much immediate effect on the price. This high volume of trading does not mean that volatility is absorbed; liquidity (availability of many buyers and sellers) is not related directed to volatility (market risk), but traders like liquidity in the market as an issue separate from volatility levels. In an illiquid market, buyers and sellers do not necessarily find one another without much searching and without substantial price movement. Energy is the most liquid commodity, and it is traded in many forms, including futures and options, stocks of energy companies, index funds, and exchange-traded funds. In other words, any market with an abundant supply of willing buyers and sellers is a robust and healthy, liquid market.

Another feature of a high-volume, liquid market is price change. Once a price has been set in the market by trading activity and agreement between buyers and sellers, liquid assets like oil tend to trade close to that latest price. Thus, liquidity may also be a reference to reliability or to the slow movement of price. The available number of units (shares, contracts, or units) on the market is called market depth, and in the energy market there is adequate interest in trading to ensure continuous high liquidity.

In comparison, an illiquid market does not readily encourage trading due to one of two causes: either the value of the asset is not widely known, or the market is thin or nonexistent (meaning buyers cannot easily locate sellers or vice versa). In the stock market, even highly liquid stocks may experience illiquidity when institutions attempt to trade large blocks of stock in single trades. The subprime mortgage market became highly illiquid after 2008 due to the uncertainty about mortgage values even for secured debt. In the oil industry, *value* is well known due to the orderly markets in commodities and stocks, but the value of these assets is easily affected by perceptions of future changes in supply.

Key point: In the commodities market, *perception* of supply and demand has as much impact on prices as do actual supply and demand.

The volume of trading itself is used as one definition of liquidity. The volume of trading in stock and commodity markets is clearly more fluid than in markets for real estate, for example; by this definition, the public auction markets are far more liquid than real estate. A related definition of liquidity then is "unit price." A share of stock at $75 per share is far more liquid than a single-family home priced at $275,000. The unit price affects the frequency of trades, so that liquidity becomes a factor of price as well as of the frequency of trades.

Liquidity is also affected by speculation in the market. It is not certain whether speculation is a reaction to changes in liquidity or its cause. On May 32, 2012, Bill O'Reilly stated on the Fox News show *The O'Reilly Factor* that oil speculators are "crooks" and should be assessed a 50% fee on their profits: "They don't want the oil. If they want to go to Vegas, go [to Vegas].... I don't want a commodity that my family needs to live to be the subject of speculators who are gambling like it would be in Vegas."[11]

On the other hand, several facts show that speculation often occurs as prices have declined, not the opposite. A statistic analysis shows that

> the so-called evidence of speculation increasing oil prices is just selected data that show a positive correlation between the number of "noncommercial" traders in the oil futures markets and the price of oil. These researchers have shown flaws with the statistical analysis and noted that other markets—and the oil market at other times—have seen the opposite correlation. The most recent counterexample is the ramped up activity of natural gas traders as the price of gas has plummeted.[12]

The debate is far from settled; each side fervently believes in its position. Speculators try to exploit movement in the price of commodities and stocks to make a fast profit, but this does not prove that their activity moves the price itself. Speculation provides a form of liquidity in the market and could move prices, but no one has offered any proof that this occurs in the energy market or elsewhere. Facilitating liquidity is not the same thing as causing negative price movement; in fact, nonspeculative interests may benefit from speculation based on the history of price movement. Speculators often have been wrong in their estimates of price direction, especially in the energy market:

> The price of Brent Crude—which is the benchmark crude these days due to a glut of oil trapped in the Midwest (more on that in another post soon)—is 5% lower for delivery a year from now and 9% lower for delivery two years from now. Delaying production of oil looks like a very bad investment.[13]

The Bad Thing about Price Volatility

A volatile market is also a market with higher risk. Volatility is another name for risk (and also for opportunity), and most consumers experience price volatility in the form of higher prices at the gas pump and on utility bills.

Investors experience volatility in a much different manner. Whether you are an investor or a speculator, you know that a high-volatility product (stock, futures contract, exchange-traded fund, or option) is going to experience fast and more extreme price movement than a low-volatility product.

> **Tip:** By definition, highly volatile trades expose you to potentially greater profits *and* potentially greater losses. Before investing in a volatile product, make sure the risks are appropriate for you, given your experience, income, assets, and basic risk tolerance.

While investors are concerned about high volatility, they recognize that this is a necessary element of price appreciation. As a result, you find that when prices are moving in the desired direction, volatility is viewed as a positive force. Of course, when prices move against you, volatility translates to a negative force causing value to fall and placing your portfolio at risk. Price volatility is not truly positive or negative; it depends on the position you hold in a product.

Most investors view liquidity as an essential and desirable force in the energy market. This is the – commodity with the highest trading volume, which also means commodity index funds, ETFs, and options will be liquid, perhaps even more liquid than the commodity contract itself. No one wants to be in a low-liquidity market for the reasons already cited (difficulty finding a market, slow price movement, and the risk that a single trade will have a big impact on overall price levels). High volatility is better from an investment point of view because both buyers and sellers have a ready market. They will not always like the price of a share, contract, or unit, but it is always available for trading.

Ethanol: A Solution or a New Problem?

The energy market is diverse and influences many other markets. This is a global effect. For example, agriculture worldwide is affected by the energy market in many ways. One of the most significant is the controversial conversion of food crops into ethanol.

Fermenting sugar or corn into fuel is an attractive concept. The problem, however, is that as the portion of crops devoted to ethanol production increases, food stores decline and prices rise. In the United States corn ethanol has been a major production effort, but food prices rose as a direct result. This phenomenon—and the entire realm of biotechnology—is not a new one. Ethanol can be traced back over 9,000 years when it was fermented as an alcoholic beverage in China.[14]

Today, ethanol is most widely used as an additive to automobile fuel. Brazil has developed ethanol from sugar cane, and ranks second only to the United States as a producer. Brazil produced 5.57 billion liquid gallons of ethanol in 2011, and production in the United States totaled 13.9 billion gallons of ethanol from corn. Together, the United States and Brazil account for approximately 88% of the world's ethanol production.[15]

Brazil allows up to 25% of fuel to consist of ethanol while the ratio in the United States is 10%. Estimates of US ethanol production are based on agricultural capacity, but the troubling negative effect on food prices curtails the potential for further development. The goal is to eventually replace fossil fuels with ethanol, but the amount of land required for this is too high, and competing with food needs creates a conflict as well. A 10% blend in US autos would require over 14 billion gallons of ethanol per year based on today's consumption; however, approximately 30 billion gallons will be required if all goals are met. The three major auto manufacturers (GM, Ford, and Chrysler) have projected that by 2014 one-half of all new vehicles will be flexible fuel vehicles (FFVs). A comparison between estimated growth of ethanol production and projected demand could point to a future shortfall.[16]

Key point: Even if all of the agricultural land in the United States were converted to growing corn for ethanol, it would not yield enough to satisfy the demand—and this would cause food shortages.

Cost Factors: Refining and Production

Many consumers and speculators believe that cost is controlled in a sinister way by OPEC meeting behind closed doors in Vienna, where OPEC is headquartered. A certain amount of price control actions does come out of these meetings. But the real influence on cost comes from limitations of refining and production capacity.

These limitations affect the supply of refined products, and in fact more production does not mean more oil gets to the market. Refining is limited so that the real test of supply comes down to a question of capacity. And over a six-year period capacity has shrunk among refineries in the United States. The number of operating refineries between 2007 and 2012 is shown in figure 2.2.

The same trend was seen in the production of barrels per day in US refineries over the same period. This is summarized in figure 2.3.

Cost Effects: Weather and Accidents

Another factor affecting prices as well as demand is the weather. Bad weather, especially in areas where refining is concentrated (such as the Gulf of Mexico) translates into shortages, as for example, after Hurricane Katrina. Accidents, such as the BP oil spill, mean less

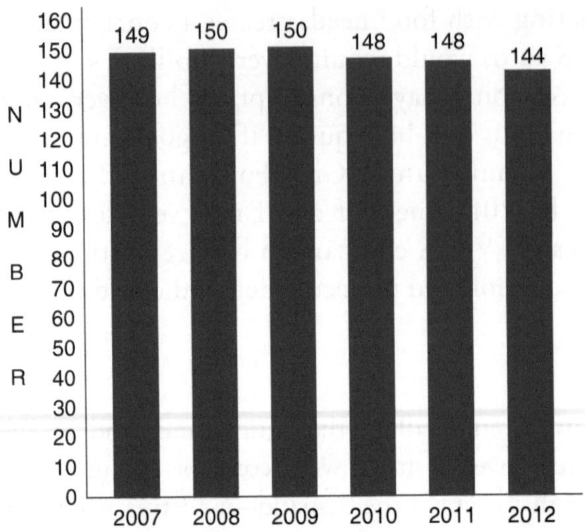

Figure 2.2 Number of operable refineries in the United States
Source: Compiled by author based on raw data atEnergy Information Administration, at http://www.eia.gov/dnav/pet/pet_pnp_cap1_dcu_nus_a.htm, June 22, 2012160150 140130120110100

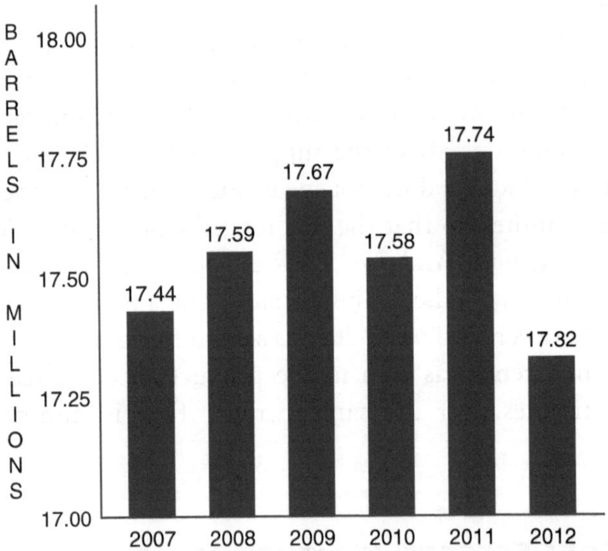

Figure 2.3 Capacity, barrels per day for operable refineries in the United States
Source: Compiled by author based on raw data atEnergy Information Administration, at http://www.eia.gov/dnav/pet/pet_pnp_cap1_dcu_nus_a.htm, June 22, 2012

deep-water drilling as potential environmental disasters slow down production.

These many influences on supply are crucial given the widespread applications of energy products. The next chapter examines the uses of these products in many industries.

CHAPTER 3

THE MANY USES OF ENERGY PRODUCTS

> Let me tell you something that we Israelis have against Moses. He took us 40 years through the desert in order to bring us to the one spot in the Middle East that has no oil.
>
> Golda Meir, in *New York Times*, June 10, 1973

The well-known uses of energy products only scratch the surface. In fact, humans rely on energy for virtually every aspect of their lives, for a vast number of products, for manufacturing, medicine, and even for processing food.

THE BEST-KNOWN USES: FUEL, INDUSTRIAL POWER, HEATING, AND COOLING

Few things work without energy. Manufacturing, production, transportation—these are the basics of all things. How do products get into stores? Without energy, they don't get there. Factories, businesses, and homes rely on energy for all types of heat and cooling. It's a huge industry because we all need it. The major energy subsectors are petroleum, coal, natural gas, renewables, nuclear energy, and electricity.

Key point: Virtually every aspect of the economy and commerce depends on energy products in one form or another and usually on more than one.

PETROLEUM

The fundamentals of petroleum demonstrate how indispensable these products are in everyday life. Oil and gas represent the best-known of petroleum products. Everyone relies on petroleum for transportation, heating and air conditioning, and electricity.

Beyond this, petroleum is used in manufacture of feedstock, chemicals, and plastics. Three-quarters of all petroleum refined is used for gasoline, heating oil, diesel fuel, and jet fuel. However, there is much more to petroleum. For 2012, the breakdown of petroleum uses is summarized in figure 3.1.

Among "all other" uses are petrochemical feedstocks, petroleum coke, heavy fuel oil, asphalt, various lubricants, waxes, and kerosene. Petroleum is used in many industries, not only in transportation-related ones but even agriculture.

COAL

Coal is widely used to generate electricity, and the worldwide coal reserves are abundant. However, the challenge is in creating clean coal technology, an effort that is underway in many countries. Coal

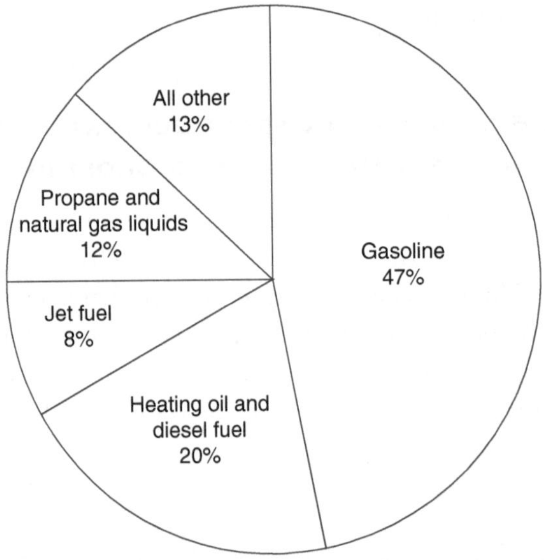

Figure 3.1 Petroleum products, 2012

Source: Prepared by author from raw data in "What are the products and uses of petroleum," U.S. Energy Information Administration (www.eia.gov), April 2, 2013

in its commonly used form is a pollutant; however, since the start of the twenty-first century, the use of coal has grown more than that of any other source. Approximately 6.6 billion tonnes (metric tons, or 2,204.6 pounds) of hard coal were used in 2012, and 73% of the world's coal output is consumed by China, the United States, and India, as shown in figure 3.2.[1]

Investors tracking fundamentals of the coal industry face a dilemma. On the one hand, coal is abundant, and on the other hand it is essential in a wide array of products. In the future, the most important test of coal as an investment possibility will be focused on liquefaction and clean coal technology.

Coal liquefaction is the process of converting solid coal into liquid, which makes pipeline transport possible and thus reduces the cost of moving coal from mines to cities. China, the largest coal consumer in the world, also has large coal reserves—and yet, China imports coal from Australia and Indonesia. The distance between coal mines and the industrialized areas of China makes rail transport expensive; increased handling capabilities in ports have also made importing coal economical for China.

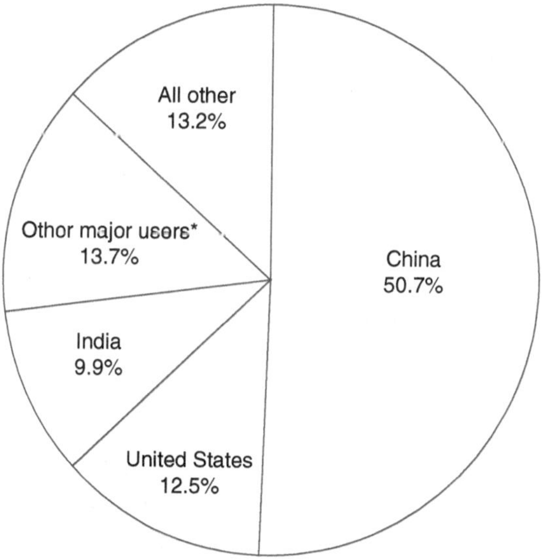

Figure 3.2 Annual coal consumption
*Russia, Germany, South Africa, Japan and Poland.
Source: Prepared by author from raw data in "Total coal consumption," U.S. Energy Information Administration (www.eia.gov), through 2011

Key point: Even though China has the world's largest coal resources, it imports coal—because importing is cheaper than transporting coal from the mines to the industrial areas of the country. Coal liquefaction would solve this problem with the use of pipelines.

The cost of transporting coal would be greatly reduced through liquefaction. There are two technological versions of liquefaction, direct and indirect. The direct form involves conversion from solid to liquid directly; under indirect methods, solid coal is mixed with carbon monoxide and hydrogen through a gasification process to create liquid hydrocarbons. Both processes involve high levels of CO_2 emissions. An alternative, synthetic coal liquefaction, is cleaner.

Clean coal technology is intended to reduce the negative environmental impact of burning coal. These technologies are designed to remove pollutants and impurities from coal before it is burned. From an investment point of view, this new technology may provide attractive and economical alternatives to traditional and highly polluting coal use methods, thus presenting opportunities for investors. However, both liquefaction and clean coal technology are far from perfected and may become viable only in the future.

NATURAL GAS

Estimates are that reserves of natural gas in the United States are massive. As a replacement auto fuel, natural gas could solve the problems of supply and environmental impact. In the nineteenth century, natural gas was used to light street lamps and almost nothing else, but thanks to today's advanced capabilities, the uses of natural gas have expanded.

In fact, natural gas is cheap and plentiful, and investment in its future may be focused on conversion of autos to use this as a fuel source. Natural gas is used in many applications, including industrial and commercial, and residential. (The term "industrial" refers to manufacturing, assembly, and warehousing, and "commercial" includes a range of income-producing activities.) The breakdown is summarized in figure 3.3.

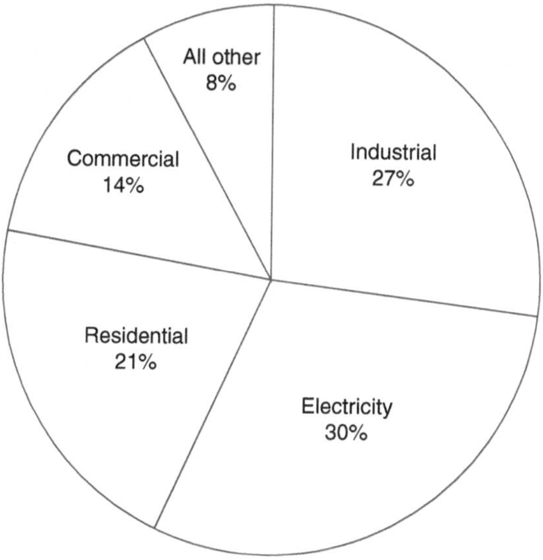

Figure 3.3 Natural gas consumption by sector
Source: Prepared by author from raw data in "Natural Gas Monthly," U.S. Energy Information Administration (www.eia.gov), as of 2010

The idea of a car powered by natural gas should intrigue investors. Although reserves of natural gas are abundant in the United States, conversion of autos to burning natural gas is going to demand many years of research and development and a lot of investment. But this is not a new idea; in 2010 14 million cars were run by natural gas worldwide, and by 2012 that number totaled over 15 million.[2]

> **Tip:** Natural gas is possibly the most likely alternative auto fuel of the future. It has been applied already and used in millions of cars.

As many as 120,000 of these cars were operated in the United States and 1,000 fueling stations are in operation, mostly in urban areas (and only about half are available to the public as of 2013). As the number of fueling stations grows, natural gas will be accepted as an attractive alternative energy source, costing as much as $2 less per gallon than gasoline.[3]

In terms of environmental impact, natural gas is a desirable fuel. Natural gas power is cleaner than gasoline and reduces pollutants significantly. Reductions made possible by newer natural gas vehicles (NGVs) reduce carbon monoxide by 70–90%, nonmethane organic gas by 50–75%, nitrogen oxides by 75–95%, and carbon dioxide by 20–30%.[4]

The trend toward using cheaper, cleaner, and more abundant natural gas points to great potential for investment in companies developing conversion of gas-powered vehicles, fuel stations, parts and service, and new cars powered entirely by natural gas. Developing vehicles and fueling stations will become a major investment with growth potential. Investors may easily overlook natural gas because it is so plentiful, but the investment opportunity will be based on the use of gas by vehicles rather than on the resource itself.

Key point: There is a double benefit to developing natural gas as an automotive fuel. It is both abundant and clean.

Among the advantages of natural gas is also safety. Natural gas is lighter than air, so in an accident the gas dissipates into the air. In comparison, gasoline is more likely to catch on fire and cause further injury and death. But there are disadvantages as well. Cars powered by natural gas tend to be slower in acceleration than gasoline-powered cars, but the difference is not substantial enough to eliminate natural gas as an alternative power source.[5]

This is a new market. The obvious problem is that only a limited number of fueling stations are in operation; however, it is possible to fill up the tank by tapping into home sources of natural gas, which in the future could revolutionize the way drivers fill up their tanks. As of 2013, only one vehicle (the Honda Civic Natural Gas) was on the market, but plans for further development are underway at GM and Ford for autos that run entirely on natural gas or on both gasoline and natural gas. The fueling issue is a big problem, however. Here again, investment in the future of natural gas may focus on this as a subsector. Home refueling could make the traditional service station obsolete. However, home refueling is far off; today, a refueling device for home use costs as much as $5,000.

RENEWABLES

Renewable energy includes many forms: solar, wind, geothermal, and hydro energy are among these. Approximately 16% of global energy is generated from renewable sources, consisting mostly of biomass.[6]

Much of the interest in expanding renewable energy sources is derived from the finite supply of fossil fuels, coupled with concerns over climate change and other environmental interests. This explains why wind power is growing at the rate of 30% per year worldwide. A 2011 projection by the IEA estimated that within 50 years most of the world's electricity could come from solar power generators, replacing reliance on coal, which is still dominating today.[7]

> **Key point:** The appeal of renewables such as wind and solar is twofold. Both are virtually unlimited resources as well as environmentally clean. The problem for alternative fuels has been reliability versus cost and whether projections are realistic. If they are, investment in renewables could be the way to go in the future.

Investment in renewable energy will be worthwhile as long as this path of development continues. Today's reliance on fossil fuels is evolving into a new reliance on renewable sources of energy, and this points to a likely trend toward development of ways of power generation other than oil refining and coal mining. Four specific themes emerge in the renewable energy industry:[8]

1. Power generation. Today, renewables provide 19% of power worldwide. Renewables represent 100% of power in Iceland, 98% in Norway, 86% in Brazil, 65% in Zealand, 62% in Austria, and 54% in Sweden.
2. Heating. China produces 70% of the world's solar hot water usage, mostly installed in apartment complexes and providing hot water to over 50 million households. Additional heating is produced with biomass energy and direct geothermal systems.
3. Transportation. Over 24.5 billion gallons of biofuels are produced annually, displacing previous consumption of 18 billion liters of gasoline. This represents 5% of worldwide gasoline use.

Strong public support for renewable energy helps continue the expansion of this type of energy. With over 30 nations relying on renewables for 20% or more if their energy use, investment enthusiasm is growing. A 2012 public opinion survey concluded that people strongly supported renewable, such as solar and wind power, even if it meant increasing energy costs, and they favored offering tax incentives for further research and development.[9]

This is encouraging for investors, because these elements—support and a willingness to pay higher costs—aid in the development of new systems and processes.

NUCLEAR

Nuclear power is used to generate heat and electricity and provides nearly 6% of worldwide energy. This comes from 439 operational nuclear power reactors in 31 countries, as is summarized in figure 3.4.[10]

Due to safety concerns, nuclear power is out of favor in many countries. Although construction of new power plants in the United States has slowed to an estimated 12 currently planned, half of existing reactors have had their licenses extended 60 years as of 2008.[11] Following the Fukushima Daiichi nuclear event in Japan in 2011, many countries altered their plans for additional nuclear power. Germany plans to close all of its nuclear reactors by 2022, and nuclear power has been banned in Italy. Estimates of future generating capacity have been cut in half by the International Energy Agency (IEA).[12]

> **Tip:** Future investment in nuclear power could be very profitable because the industry has an excellent safety record, especially compared to oil and gas and the coal industries. But perception has prevented expansion of nuclear power in many countries, including the United States.

Although disasters have made nuclear power less popular than other forms of energy, it is one important source of power generation. Investors may benefit from the reduced number of newly built plants, because scarcity works in favor of companies and investors pursuing less popular forms of energy.

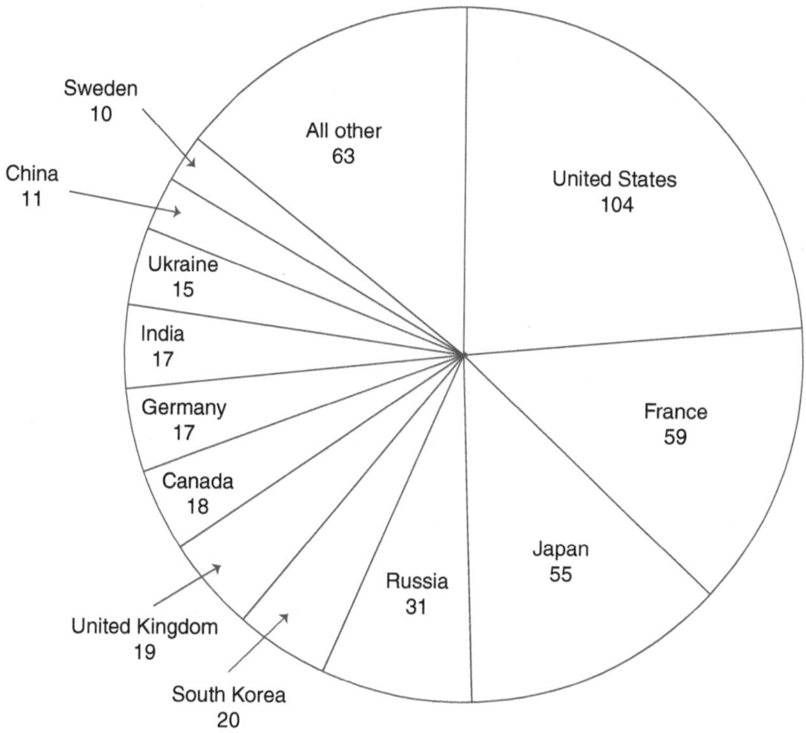

Figure 3.4 Operable nuclear reactors worldwide
Source: Prepared by author from raw data from World Nuclear Association (www.world-nuclear.org) as of 2010

ELECTRICITY

Electricity has a wide array of uses. Since the invention of the lightbulb in the late nineteenth century, electricity has been a primary source for many types of power, residential and commercial. The three largest users of electricity are the United States, China, and India. However, a comparison between population and electricity use reveals a disparity, caused by lifestyle as well as by production capacity as illustrated in table 3.1.

The basis of electrical power is the use of transistors, which are used in all modern circuitry. Today an integrated circuit is likely to consist of several billion miniature transistors in relatively small devices only a few centimeters thick.[13]

This is significant because growing numbers of automated processes define how the modern world works, not only in manufacturing but also in information technology, communications, transportation, finance, and virtually every other industry as well. Electricity, a relatively simple force of nature, has been harnessed to make possible big advances in technology and the use of energy.

Tip: Electricity is unlimited as a force of nature. It is harnessed in numerous ways, and the combination of electricity with the transistor has already revolutionized many industries.

Energy is created primarily with coal, both in the United States and worldwide, and secondarily with natural gas. The sources of electricity in the United States and worldwide are shown in figure 3.5.

Table 3.1 Electricity usage, China, India, and the United States

Country	**% of world Population (bil)**	**Electricity use**
China	1.34	17%
India	1.17	4
United States	0.31	22
All others	4.26	57

Source: "IEA Statistics/Electricity and Heat by Country," www.iea.org, 2009.

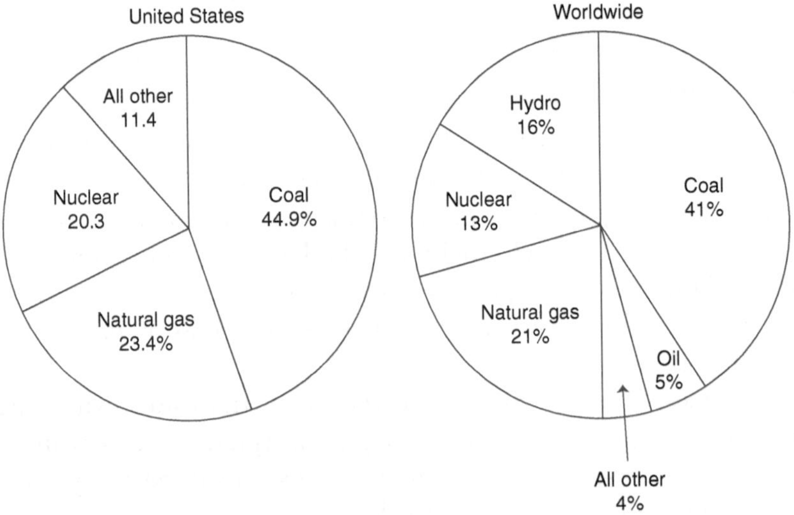

Figure 3.5 Sources of electrical power
Source: Prepared by author from raw data from Energy Administration (www.eia.gov), Electric/Heat, December 2012 and Electric Power Monthly, March 22, 2013

LESS OBVIOUS USES: HOW OIL PRODUCTS ARE BROKEN DOWN

Oil products provide much more than power for autos, home utilities, and manufacturing. Among the many uses for petroleum products are:

Propane and other fuels.
Blended fuels for various grades of gasoline, kerosene, aviation fuel, and diesel—with additives including dyes, antiknock additives, and antifungal compounds.
Sulfur used in industrial materials.
Wax for packaging of frozen foods and other products.
Lubricants such as machine oil, motor oil, and grease.
Tar for packaging and roofing.
Asphalt for gravel binding and concrete used to pave roads.
Petrochemicals and feedstock, also developed for use in manufacturing plastic products, fibers, solvents, and chemical compounds.

The next chapter moves beyond the fundamentals introduced in the first three chapters, and explains the opportunities and risks in the energy market.

CHAPTER 4

OPPORTUNITIES AND RISKS

> The key to success in the risk/return game is not to minimize or maximize risk, but to manage it.
>
> Mark A. Johnson, *The Random Walk and Beyond*, 1988

IN THE ENERGY MARKET, AS IN ANY OTHER MARKET, THE GREATEST difficulty investors face is understanding that opportunity and risk are opposite sides of the same coin. The greater the opportunity in energy investments, the greater the risk must be.

Even if you have heard this before, it remains true that investors expose themselves to excessive risk unintentionally. This happens when the focus is on the opportunity alone. Successful investing relies on (a) experience in the market, (b) knowing your risk tolerance, (c) studying the fundamentals, and (d) matching risk and product appropriately.

Key point: You cannot escape risk in any investment. The more promising the opportunity to earn profits, the higher the risk. But knowing this and acting on it are two different things.

Experience in the market means having capital at risk, going through the experience of buying and selling, and ending up with profits *and* with losses. As difficult as it is to accept losses, you might learn more from them than you learn from profits. "Experience is the name everyone gives to their mistakes," as it has been said.[1]

Knowing your risk tolerance is not just a matter of identifying levels of risk to various investment decisions, but also of knowing how much risk you (a) can afford to take and (b) are willing to take. This process of defining risk tolerance is not simple and encompasses your knowledge of the risk universe, age, income, experience, family status, income, assets, and perceptions of markets. For example, if you believe that all energy investing carries a high risk and is likely to lead to big losses, then your risk tolerance should steer you away from energy. If you think your belief might be based on assumptions that are not always true, then you may expand your risk tolerance by further research.

Studying the fundamentals involves not only the well-known art of reading financial statements and tracking ratios and financial trends. It goes beyond that. The fundamentals in any industry should include gaining knowledge about the markets, supply and demand, competition, and other attributes about the product and range of products involved. For example, the energy market is dominated by the integrated oil and gas industry, but there is much more to it. The fundamentals help you to expand your knowledge to determine if and when investing in this market is sensible.

Matching risk and product appropriately requires thorough analysis and understanding of the attributes of a product and the market for it. For example, you might reject investing in a coal mining company based on plentiful supply, low cost, and inherent environmental problems. All of these are valid concerns; however, a match between risk and product might expand to an investigation of coal liquefaction or clean coal technology. A company attempting to advance these markets could be a strong investment candidate, compared to the more obvious investment directly in coal mining.

Energy Markets, Incredible Opportunities

As such a large industry (or more accurately, range of industries) the energy markets are vast. As a basic necessity for everything humans produce and consume, the energy market also produces jobs and economic growth in every other sector.

The energy industry consists of the following subsectors:

Light crude oil, distinguished by low wax (and "sweet" crude has low sulfur content). These low content characteristics aid in refining oil. Crude is refined into other products: gasoline, kerosene, and diesel gas. A major classification of light sweet crude is Brent crude oil. It consists of oil from the North Sea and is a worldwide benchmark for pricing of oils from the Atlantic basin.[2]

In addition to the oil fields drilled traditionally, huge reserves of oil are located worldwide in shale. In the United States there are two large fields of shale, the Devonian-Mississippian shale in the Appalachian Basin and the Green River shale formation in parts of Colorado, Utah, and Wyoming. The oil reserves of the United States are second only to China's. The US reserves alone are estimated to be four times greater than proven oil reserves in Saudi Arabia.[3]

Tip: The US oil reserves are massive, even greater than those of Saudi Arabia. But many reserves cannot be tapped due to location or environmental challenges.

The shale resource worldwide is depicted in figure 4.1.

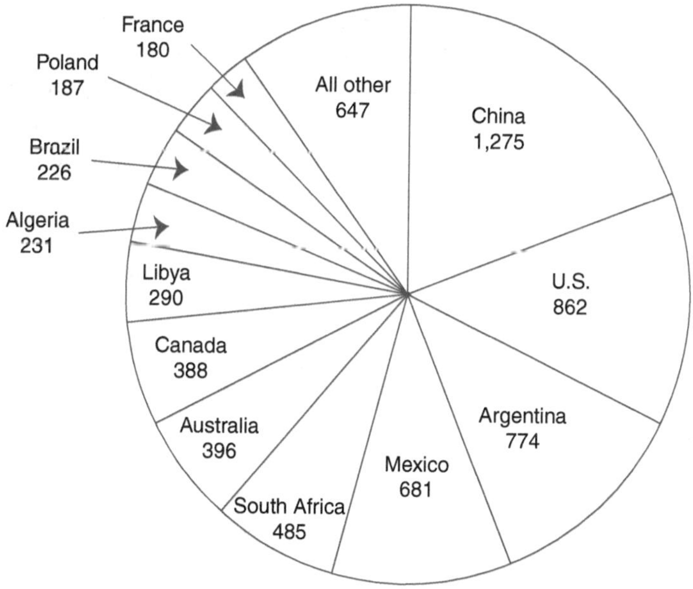

Figure 4.1 World recoverable shale gas
Source: Prepared by author from raw data in "World Shale Gas Re*Sources*," U.S. Energy Information Administration (www.eia.gov), April 5, 2011

Investment opportunities in light crude prices are focused on the futures market. This market is accessed by direct trading in futures or, for most people, in a less expensive method using exchange-traded funds (ETFs) or commodity index funds. The debate over supply and reliance on foreign oil fuels both the supply/demand issue and speculation in oil.

Heating oil is derived through the refining process. About one-fourth of crude oil ends up as heating oil, which is the most widely used source of heat in the Northeast of the United States. Although the futures prices of heating oil are likely to track oil prices in general, severely cold winters will drive up demand and prices will follow.

Ethanol is an additive, but it relies on large amounts of agricultural product. In the United States, the experiment with ethanol has been based on corn (in Brazil, sugar cane was the product of choice). The problem with devoting corn to the production of gasoline additives is that this raises the price of food. However, there is not enough agricultural acreage in the United States to meet the minimum needs for ethanol:

> Ethanol land requirements: Approximately 50 gallons of ethanol are produced per acre of corn. Thus 2.8 billion acres of land would be required to generate the 140 billion gallons of fuel used in the USA annually, which is more than 5 times all of the cropland that is actually and potentially available for all crops in the USA.[4]

Unleaded gasoline is well-known as the primary auto fuel worldwide and accounts for approximately one-half of oil consumption in the United States. The futures contract for unleaded gas is labeled RBOB (Reformulated Blendstock for Oxygen Blending). This is the most liquid form of refined oil.

Everyone who visits the gas station knows that prices of unleaded gasoline are volatile. However, as an industry, this sector of "oil and gas" is very diverse and encompasses all phases of the process: exploration, drilling, refining, transportation, and even service stations. Other industries (such as autos, transportation, retail, and manufacturing) also rely on gasoline to complete their supply chain and to bring goods to the market. As the price of gasoline changes, so do prices of other commodities.

Tip: When the price of gasoline rises, so does the price of everything people buy in stores. The cost of transporting goods to the stores is dominated more than anything else by the cost of fuel. That is, pump price rises are only the beginning of the story.

The market for unleaded gasoline relies on refinery capacity more than on supply and demand. Refineries produced a lower level of refined product in 2012 than in the four preceding years (see figure 2.3 in chapter 2). Thus, supply and demand mean less in the market than what refineries are able to bring to market. If demand grows but the number of refineries remains fixed or drops, prices will rise. Investors may consider investing in refinery production via stock ownership or ETF investing rather than by speculating on future prices of unleaded gasoline. A list of refineries, dates built, and capacity is found in table 4.1.[5]

Table 4.1 US refineries: Years built and capacity

Year Built	First Operated	Location	Original Owner	Original Capacity	Current Owner	2011 Capacity
2008	2008	Douglas, WY	Interline Resources	3,000	Garco Energy	3,600
1998	1998	Atmore, AL	Goodway	4,100	Goodway	4,100
1993	1993	Valdez, AK	Petro Star	26,300	Petro Star	55,000
1991	1992	Eagle Springs, NV	Petro Source	7,000	Foreland	2,000
1986	1987	North Pole, AK	Petro Star	6,700	Petro Star	19,700
1985	1986	Anchorage, AK	ARCO	12,000	ConocoPhillips	15,000
1981	1982	Thomas, OK	OK Refining	10,700	Ventura	12,000
1979	1980	Wilmington, CA	Huntway	5,400	Valero	6,300
1978	1979	Vicksburg, MS	Ergon	10,000	Ergon	23,000
1978	1979	North Slope, AK	ARCO	13,000	BP Exp AK	12,780
1978	1978	North Pole, AK	Earth Resources	22,600	Flint Hills	219,500
1977	1978	Lake Charles, LA	Calcasieu	6,500	Calcasieu	78,000
1976	1977	Garyville, LA	Marathon	200,000	Marathon	464,000
1976	1977	Krotz Springs, LA	Gold King	5,000	Alon	80,000
1975	1975	Corpus Christi, TX	Saber	15,000	Valero	142,000
1967	1967	Good Hope, LA	Kirby Industries	6,500	Valero	205,000

Source: U.S. Energy Information Association, Refinery Capacity Report, July 18, 2013, at http://www.eia.gov/tools/faqs/faq.cfm?id=29&t=6.

Refining affects all petroleum products; however, due to the high use of unleaded gasoline (about one-half of total oil consumption in the United States), the effect on prices of refining capacity and future expansion as a factor in price trends is significant.

Natural gas is favored for environmental reasons because it is cleaner than gasoline. When burned, it generates 29% carbon dioxide less than oil and 44% less than coal. The comparison is illustrated in figure 4.2.

Natural gas is also abundant, and US reserves are 9 trillion cubic meters (TCM), more than those of all other countries except Russia, Iran, Turkmenistan, and Qatar. The breakdown of proven reserves for the six largest holders of reserves is summarized in figure 4.3.[6]

Natural gas is consumed primarily in power generation and for industrial purposes. Future conversion of gasoline vehicles to operation on natural gas may be a promising future investment opportunity, but such massive changes take time. Some environmental

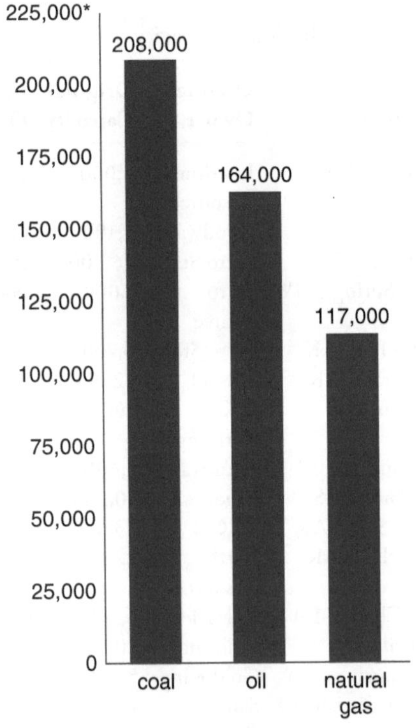

Figure 4.2 Carbon dioxide emissions
*pounds per billion BTU of energy input.
Source: Compiled by author based on raw data at Energy Information Administration (www.eia.gov),"Natural Gas Issues and Trends, 1998"

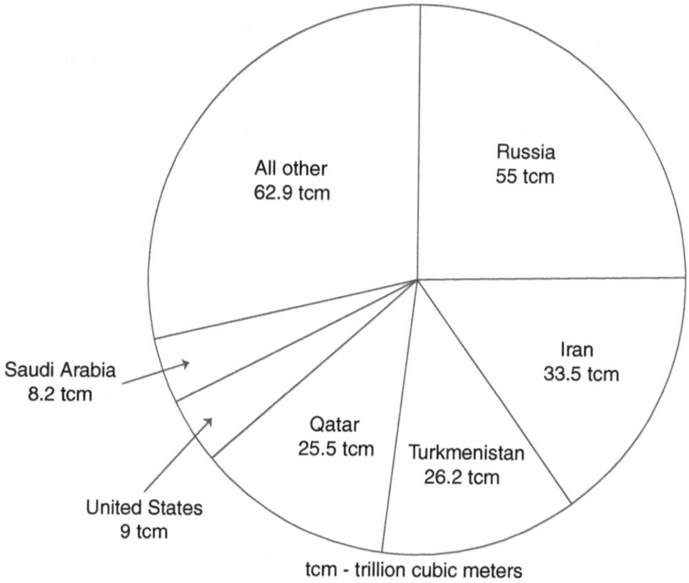

Figure 4.3 Natural gas proven reserves
Source: Prepared by author from raw data in "U.S. Crude Oil, Natural Gas, and NG Liquids Proved Reserves," U.S. Energy Information Administration (www.eia.gov), August 1, 2012 (2010 totals)

organizations are resisting the changeover. In spite of natural gas having a cleaner emissions record than oil, resistance to further mining of natural gas has slowed the development of this resource. This raises two issues for investors. Resistance to accessing of natural resources creates higher prices by artificially increasing demand, and on the other hand, the better environmental record supports future development. This is one of the reasons that conversion of automobiles is moving forward even though the future of access is debatable.

Tip: The issue of resource versus environment is an enigma. On the one hand, cleaner and more abundant fuel is desirable. On the other hand, some groups resist any and all forms of energy development on environmental grounds. For investors, this "forced scarcity" could point the way to future opportunities in the market.

Natural gas may be transported to the end user by gas pipelines. However, it can also be converted into liquid form and stored and transported in tankers as liquid petroleum gas or LPG.

Nuclear energy is a fairly safe form of energy. Incidents in the Ukraine and, more recently, in Japan have drawn attention to the dangers of nuclear contamination. There has never been a fatality in the United States due to accidents or leakage at a nuclear power plant.[7]

There have been accidents on nuclear submarines and in medical radiology errors, but as a power source, nuclear energy is far safer than other sources. For example, the Monongah Mining Disaster in 1907 killed 362 workers in an underground explosion.[8] On average, from 1880 to 1910 there were over 1,000 deaths per year in mines, but the number was reduced to an average of 500 from the 1950s to the 1990s.[9]

The reputation of nuclear power as dangerous and toxic has been overstated, at least in the United States. Compared to other industries, the nuclear track record of zero deaths makes this the safest energy sector. The number of fatal occurrences in oil and gas and other mining industries is summarized in figure 4.4.

From an investor's point of view, a problem with nuclear energy is its perception as dangerous. Therefore, any future steps to ensure the safety of nuclear power plants may present opportunities. The most recent disaster in Japan points up, however, that even with excellent safety protocols an event such as an earthquake or tsunami makes any form of energy production unpredictable. The potential contamination from nuclear power plants is a big factor in inhibiting future investment potential in this form of energy.

Alternative energy includes many variations: solar, wind, geothermal, and hydro are among these. While the energy in these sources is substantial and unlimited, harnessing it has been shown to be much more difficult than some have thought. Investors should be aware of past experiments and how they ended up.

Key point: During the 1970s when the Middle East oil embargo drove prices up, great interest arose for solar, wind, and other alternative forms of energy. These proved to be expensive and inefficient. Investors may benefit not from buying shares of alternative energy companies, but from investing in development of more efficient technology.

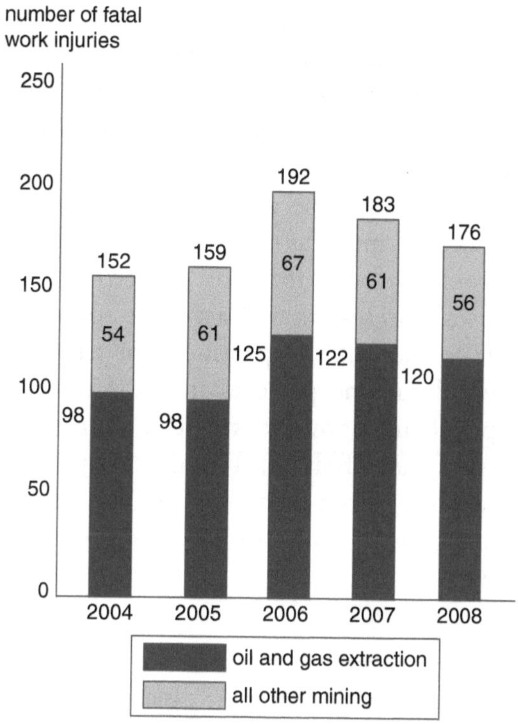

Figure 4.4 Fatal occupational injuries in the private mining industry, 2004–2008

Source: Prepared by author from raw data from U.S. Bureau of Labor Statistics, U.S. Department of Labor, 2010

Whenever the price of crude oil rises, alternative energy becomes popular and draws in investors. This does not mean the ideas are poor, but as investments they take a long time to become profitable is highly questionable. "Green technology"—that is, any system that will not pollute the environment and relies on natural sources such as sunlight—is popular as an idea among environmental groups and anyone opposed to the use of fossil fuels. However, in its long history, solar energy has not been able to surpass the efficiency and effectiveness of fossil fuels.

The problem with solar energy is the relationship between efficiency of solar storage cells and the high production costs. In the 1970s Exxon Corporation developed the first efficient solar panel that was cost-efficient, a first in the industry. Between 1986 and 1999, large-scale solar power plants were developed, and the industry appeared promising. However, since 2010, many solar energy companies have gone through bankruptcy. The markets were simply not available for the products they created. By 2012, a record number of solar plants had opened including the largest in the world, the Golmud Solar Park in China, which has a capacity of 200 megawatts. While not as large, the collective power of India's Gujaret Solar Park produces as much as 605 megawatts.[10]

A similar history is found for wind power. Whenever the price of crude oil rises, interest in wind power soon follows. In the 1970s and 1980s the U.S. Department of Energy opened and operated 13 wind turbine projects in the United States. Wind farms have opened in many areas of the country, both on land and in waterways just offshore. The disparity between *capacity* and average output is often great, and many wind power experiments have failed as a result of not producing the predicted amount of power.

Tip: For alternative power sources, capacity defines maximum output. However, capacity is almost never reached, and even when it is, the level is temporary. Wind power, for example, is unreliable because the wind is not always blowing.

Hydroelectricity may be the most promising of all alternative energy sources because it is the cheapest and easiest to produce. Several recovery methods are in use. These include dams as the most productive. In operation today, the largest hydroelectric dam is the Three Gorges Dam in China, with a capacity of 20,000 megawatts.

A second type is called pumped-storage hydroelectricity and requires construction of a series of reservoirs. When demand is low, excess capacity is pumped to higher reservoirs. When demand increases, water is released to lower reservoirs where turbines and generators generate electricity.

The "run of the river" method consists of electric power stations, and electricity is generated from the power of water along the river's course. In 2011, these systems provided nearly 12% of electricity in the United States.[11]

Systems relying on tidal power generate energy from the movement of water as tides ebb and flow. These are very predictable, but only a limited number of sites are suitable for power plants.

Underground power stations take advantage of height differences between waterfalls and rivers or lakes. In these systems, gravity is used to generate electricity.

All of these hydroelectric sources rely on the affordability of plants construction and operation compared to the actual electricity generated. The often-cited capacity of a system is electricity generated at peak performance; few plants operate at this peak all of the time, and actual output actually may be substantially lower than capacity.

Investors pursuing alternative energy via corporations constructing plants or through commodity futures should be aware of the history of these programs. They carry a high risk if only because the costs are high and sustaining a plant to generate more power than it costs to produce has been shown to be elusive in the past. If a program relies on energy prices continuing to rise to make the alternative power source viable, the plan could be in trouble. As more fossil fuel reserves are discovered and as advances in natural gas technology continue, the investment prospects for alternative energy are becoming less certain.

Liquid hydrogen and clean coal are energy sources that have captured the imagination of energy companies and also of those interested in preserving the environment. Liquid hydrogen is potentially in endless supply and its generation and use carries no environmental damage. The only by-product of generating liquid hydrogen is water. Today, liquid hydrogen is used as a rocket fuel, but future technology may also improve so that it will be practical as auto fuel. For investors, the technology has potential to work. There is no scarcity issue for liquid hydrogen itself; the value for future investment is in the technology and conversion of autos and other vehicles.

Key point: Like many forms of alternative energy, liquid hydrogen itself is in unlimited supply. The cost is found in the technology to develop the fuel and modify vehicles so it can be used. Future investment in this technology offers great potential, but the same question keeps coming up. It's not how but when that matters.

Given the zero emissions possible with liquid hydrogen as fuel, the technology may solve many problems, eliminating fossil fuel scarcity as well as ensuring environmental safety. However, liquid hydrogen has to be stored in thermally insulated containers requiring careful handling because it is highly flammable and may also dissipate over time. These concerns could be overcome in a cost-efficient way if liquid hydrogen were to be produced for widespread use as a fuel for vehicles.

The goal of producing clean coal is a controversial idea. Some people and organizations believe it is a fantasy, and others see coal liquefaction as a new industry with great potential. Environmental groups have actively promoted their belief that there can be no such thing as clean coal.[12]

However, politicians on both sides of the political spectrum have expressed support for clean coal technology and its potential. The American Reinvestment and Recovery Act of 2009 included funding of $3.4 billion for research into carbon capture and storage for clean coal technology.[13] Cabinet members further expressed support for the technology, including Energy Secretary Steven Chu.[14]

The cost of transporting solid coal is so high that liquefaction and the use of pipelines could revolutionize the industry. Coal is plentiful; if clean coal is a future reality, it could solve many problems. The future of investing in this technology depends on how pollutants are managed. Some clean coal technology removes pollutants, but other technologies actually create more pollution than are produced with refining crude oil.

Investment Risks and Opportunities

As with any market, energy presents fundamental risks and opportunities in several areas. These include inflation, market volatility, cross-sector considerations, diversification, and individual risk tolerance.

Inflation is poorly understood even among experienced investors. In fact, it is a reflection of lost purchasing power of money. A widespread and persistent belief is that inflation is *caused* by prices going up when in fact the price increase is the consequence of inflation, not its cause. Inflation is measured by rising prices, so it is easy to understand why this belief persists. Putting this another way, "Steel prices cause inflation like wet sidewalks cause rain."[15]

Market volatility is experienced when prices move rapidly or in ways you cannot predict easily. The degree of volatility defines market risk. But you can reduce that risk in many ways, such as by buying mutual funds or ETFs in the energy sector, buying stock in companies that do more than just explore or drill, and investing in companies that tend to not move in the same direction as energy companies (hedging).

Cross-sector considerations refer to energy's direct influence on most other market sectors. For example, if a high percentage of the corn crop is used to produce ethanol, food prices will rise. In this respect, energy demand affects food prices. If energy costs rise, so do transportation costs and, consequently, all retail prices rise as well. It is difficult to imagine a scenario in which trends in energy do not affect other sectors.

Diversification is not well understood. It does not mean buying stock in three different oil companies. By doing so, you are adopting the position, "Why should I risk losing all of my money in one company when I can lose it just as efficiently in three?"

True diversification consists in spreading risks among investments that are not likely to react in the same way to changing market conditions. It can be accomplished in several ways. Among them are direct ownership of stock in several companies, including energy and other sectors. Even within the energy sector, you can diversify by owning shares in oil exploration, drilling, refineries, transportation, alternative energy, and oil equipment. These are all part of the sector but concern different aspects of it. Among the tests investors use in picking stocks are price/earnings ratio (P/E), dividend yield, revenue and net income trends, and debt ratio. Technical traders also track price volatility to identify the market risk of a particular stock.

Another way to diversify is to buy shares in commodity index funds. These are forms of mutual funds that spread a portfolio among commodity sectors including energy, agriculture, and precious metals. The index funds specify the percentage held in each type of commodity, and energy usually accounts for a majority of the overall portfolio. Yet another way to diversify is to select an exchange-traded fund (ETF). This is a mutual fund that identifies a "basket of securities" so you know in advance where the fund invests. An energy-related ETF may be quite specific to a subsector or may focus on a particular country or region.

Diversification is also possible by investing in equity (stock) or by trading options on energy stocks or ETFs within the energy sector. Options trading is appropriate for traders who understand the risks and strategies available and are able to manage an options portfolio. Those who do so can accomplish diversification without needing to increase capital at risk.

A distinction is made between diversification (spreading risk among different products) and asset allocation (spreading risk among different markets). For example, a diversified portfolio might consist of stocks in energy, retail, agriculture, finance, and technology. An asset allocation program is a division of total capital among markets including equity (stocks), debt (bonds), and real estate. All of these can include direct ownership or holding shares in index funds, ETFs, mutual funds, and similar pooled product investments.

Individual risk tolerance is just as often misunderstood as diversification. "Risk tolerance" for some investors means "the amount I can afford to lose." More accurately, though, it is a test of the opportunity level you want to pursue in exchange for the risk level that is involved. Risk tolerance is defined by experience, knowledge, long-term goals, family income and assets, and individual bias toward or away from specific types of investing and trading activity.

The energy market can be defined in terms of risk tolerance. Opportunity and risk are attributes of the same aspect of any form of investing, and every investor needs to be keenly aware of how products and risk tolerance match up. Some investors make the mistake of trading in products without thoughtful analysis, often on the basis of daily news stories, economic data, tips from friends, or changing prices in stocks or commodities. However, without a deeper understanding of the risk attributes in a particular investment, it is impossible to know whether a particular product (stock, commodity, mutual fund) fits within your individual risk tolerance level. This is perhaps the greatest error investors make in how and why they pick one investment over another, not to mention when and why they buy or sell.

This chapter and the previous three provided you with a broad overview of the basics of the energy market. The next section includes six chapters that take a deeper look at the range of energy products: crude oil, natural gas, propane, heating oil, coal, and alternative energy products.

PART II

MANY WAYS TO INVEST OR TRADE

CHAPTER 5

TRADING IN ENERGY FUTURES

> Rash indeed is he who reckons on the morrow, or haply on days beyond it; for tomorrow is not, until today is safely past.
>
> Sophocles, *Trachiniæ*, ca. 430 BC

YOU HAVE NUMEROUS CHOICES IN INVESTING AND TRADING IN THE energy market. This chapter discusses futures trading, specifically direct transactions involving futures contracts. The next three chapters explore stocks, mutual funds, and options as additional methods for getting into this market.

The selection of a particular investment vehicle determines risk in a very real way. Market risk is only one form of investment risk. You also need to think about leverage and diversification risks. In buying or selling a single crude oil future, the full contract value may be $90,000 (when oil is at $90 per barrel) because the contract increment is 1,000 barrels. However, you can leverage this up to 95% and deposit only $4,500. But this degree of leverage increases your risk. As long as the price moves up (if you buy the contract) or down (if you sell the contract), you make money on your leveraged trade. But if the price moves against you, that leverage gets expensive.

Key point: Risk is not limited to the danger of price movement. Many other forms of risk have to be considered as well, notably the risks associated with leverage and diversification.

The use of a single contract also demonstrates how difficult it will be to diversify your position. You can only trade as many futures contracts as you can afford, and even 95% leverage will cost money; if you want to diversify your risks, direct trading in futures might not be a good choice. Finally, the cost of trading has to also be considered in risk evaluation. Direct trading in futures is expensive compared to buying and selling stock or investing through the ETF or commodity index fund markets.

Risk also determines the convenience that you will enjoy in trading in a specific product area. For example, futures are considered high in risk and expensive; stocks, in comparison, are cheaper to trade but can also carry very high risk. Mutual funds are perhaps the most convenient way of all to invest in the energy sector, but profitability is not going to be as great as it will be with direct ownership of stocks. And options are complex and appropriate only for experienced traders who are able to manage risks and can accept higher than average risks.

Your individual risk tolerance defines which method is best for you, assuming that after studying the attributes of the energy sector, you have determined that it is a worthwhile industry in which to invest. If you have a lot of experience and can accept higher than average risks, you may prefer stocks, options, or even futures. If you are very cautious and conservative, you are more likely to be drawn to exchange-traded funds (ETFs) focusing on energy, to commodity index funds, or to traditional mutual funds.

No one can tell you which of these will work best for you; it's a matter of personal evaluation and finding the investments that appeal to you. The choice depends on your circumstances (experience, income, assets, and risk tolerance). This chapter begins with an explanation of the futures market and energy futures. Energy futures are traded more than any other commodity, not only because there are so many different subcategories of energy but because energy products are used in a broad range of other sectors. As a result, many companies in other sectors may trade in futures as a way to control and predict their energy costs. For example, a company in the transportation business (airlines, trucking, railroads) or freight delivery (UPS or FedEx, for example) has to contend with energy costs as the biggest influence on profitability. Therefore, these companies are inclined to track futures prices and perhaps even to trade in futures in order to control future energy costs for the services they provide.

Tip: In selecting investments, remember that energy demand is not limited to direct sales of energy products or exploration and drilling. Virtually every industry relies on the use of energy products.

This is the essential and original purpose of futures trading. Originally, farmers were able to fix the market price for their crops even before those crops had been planted, by buying a futures contract (or entering into a less formal forward contract with a buyer). Today, the same process is applied to energy futures. Companies relying on energy for the product or service they provide will want to fix the cost of fuel over coming months, and if they expect energy costs to rise, they may buy a futures contract with the intention of taking delivery of fuel at the fixed futures price, even several months from the day the contract is bought. If the price remained at or below that futures price, they would not take delivery but would be inclined to roll the contract forward to a later delivery month, or they would just take the loss and let the contract expire.

THE NATURE OF ENERGY FUTURES

A futures contract is an agreement between buyer and seller to fix the price of a commodity in the future. An energy futures contract comes about when buyer and seller agree to the price of energy several months from now. For the buyer, the reason to enter this transaction is to add certainty to prices several months away. For the seller, the transaction is more speculative; the hope is that the price will not rise, so that the seller profits by initiating a trade by selling, and later by entering a buy to close order at a lower price than the original sales price of the contract. (Buyers transact the well-known buy-hold-sell sequence, whereas sellers do the opposite, sell-hold-buy. So a buyer will initiate a "buy to open" and "sell to close" set of orders, and sellers will initiate a "sell to open" and "buy to close" set of orders.)

As long as the buyer owns that futures contract, the price of energy to be bought in the future will be fixed at the price agreed upon, no matter how much the market price changes. And the seller has agreed to provide energy products at that fixed price. The catch is that for both sides a futures contract has a limited life span and will expire at some point. That is, a futures contract has to be executed, allowed to expire, or rolled forward and kept in force. It does not exist forever, and both buyer and seller recognize that a part of the price is based on how much time remains until expiration date.

Not all futures contracts end up in delivery of the goods. Some do, but any speculative trading ends with profit or loss, not in delivery of the commodity. The old story about a commodity trader waking up to find tons of soy beans in his front yard is an exaggeration. It is easy to avoid taking delivery, and being surprised by delivery is quite unlikely. Only commercial traders actually take physical delivery of commodities; traders who do not want to take delivery (that is, most traders) simply close their trades before what is called the "first notice day." This occurs weeks in advance of the contract's expiration date, so there are no surprises. There are two ways to conclude a futures contract. The first is physical delivery, which means that the contract amount (such as 1,000 barrels of oil) is delivered and must be paid for in full. The second is cash settlement. The profit or loss is paid for by one side and received by the other at expiration. Cash settlement is applied to futures that could not be physically delivered, such as those on an index.

In a contract subject to physical delivery, taking delivery is avoided by selling the position before the delivery deadline (for long positions) or entering a buy to close (for short positions). In cases where avoiding physical delivery is necessary, it may mean having to accept a loss on the trade.

Three specific players are active in the futures market. The first is the energy consumer who needs to fix the price of energy for coming months. This includes airlines, trucking companies, and any other type of company relying on gas, diesel, and other fuel. The second player is a provider; this could be an oil company, a refinery, or a delivery service, such as the corner gas station. The third player is the speculator market. Speculators buy or sell futures contracts based on a belief about the direction in which price will move in the future.

Key point: Speculators account for much of the activity in futures trading, but the act of speculating on price does not by itself cause prices to rise or fall. The change in price is a result of supply and demand for the product.

Anyone who could possibly know what a product's price will be in a few months can get rich trading against that price, but no one can possibly know for sure in what direction a price will move, whether for commodities, stocks, or other investments. This is where the futures contract comes into the picture. Just as stock investors buy and sell shares of stock believing that price is going to move in a particular direction, futures traders also bet on price movement.

Every trader faces the same set of challenges in timing as well as in anticipating price direction. This is why you need to study both the fundamental and technical aspects of any product. The fundamentals of stocks rest with a company's financial strength, profitability, and working capital. Stock prices also depend on a company's competitive position, dividend history, and the reputation of its management. In the futures market, the fundamentals are quite different and more elusive. Several elements affect fundamental analysis of futures, including the following:

Scarcity of the product is a topic of endless debate. We are running out of oil, as many believe, but new reserves and new methods of extraction keep changing the fundamental picture. As prices rise, some methods of extraction once not feasible become profitable, which increases the supply. Domestic reliance on foreign oil has political as well as economic ramifications, and this also affects the scarcity of every petroleum product.

Expected technology advances may radically change the fundamentals. Technology making possible clean coal or liquid hydrogen as fuel sources is exciting. Hydrogen is in unlimited supply so the technological change is in how to convert it into a safe, economically produced fuel. Today, extraction methods such as horizontal drilling and sources such as sand oil and scale oil, have advanced the technological side of finding and extracting oil. However, these changes will affect futures prices not in the short term, and an active futures contract open today will not be changed by anything that might happen far into the future.

Refining limitations affect the picture as well, especially for oil and gas futures. It does not matter how much oil is drilled if the limit of refining capacity has been reached. Oil cannot be moved to the market in its pure form but must go through the refining process. There are dozens of refineries in the world, including 144 in the

United States. Of these, 43% are located in three states: Texas (29), California (17), and Louisiana (16).[1]

Most large-scale oil refineries are located elsewhere; India, Venezuela, South Korea, and Singapore all have refineries larger than any in the United States. Of course, this means that refining limitations are not domestic alone. While capacity in the United States is critically important to the domestic supply, potential problems of refining capacity are a global version of a fundamental in the energy market. The largest refineries in terms of barrels refined per day are summarized in figure 5.1.

Concerns and fears not directly related to the products themselves, such as environmental issues, are also important. Most forms of energy have harmful environmental effects, and the search for clean technology continues. However, the concerns about harmful

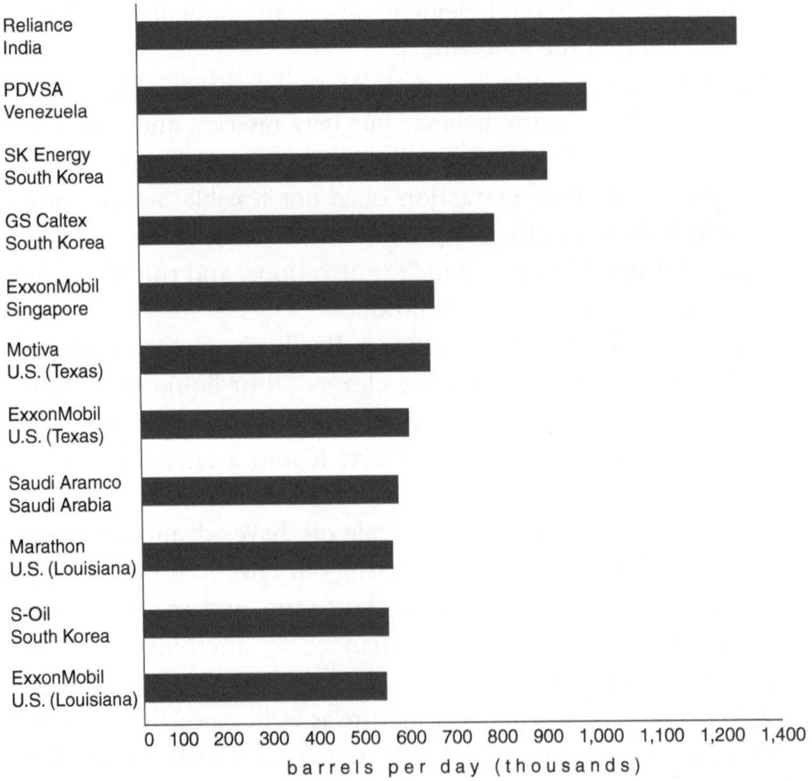

Figure 5.1 World's largest oil refineries

Source: prepared by author from raw data at "Reliance Commissions World's Biggest Refinery," The Indian Express, December 26, 2008

and toxic effects often exaggerate the threat level. For example, the nuclear industry is quite safe compared to the oil and coal industries. Even so, fear of nuclear contamination has prevented construction of new nuclear power plants in the United States even as other countries expand the number of their plants. As of January 2013, the United States had 104 active nuclear power plants or 24% of the worldwide total of 437.[2]

With these many attributes to energy fundamentals, the futures market is perhaps more difficult to analyze than the stock market. Unlike the specific financial results of a corporation for stock analysis, the energy market fundamentals include numerous economic and political factors as well. Some of the fundamentals involved are tangible and real; others grow from perceptions of supply and demand, an entirely separate factor—but one that has as much influence on price and price direction.

Tip: Futures are more difficult to analyze than stocks because energy fundamentals move well beyond the financial fundamentals associated with stocks. A futures contract cannot be studied on a balance sheet or income statement.

Futures trading is accomplished by way of a contract, which can be bought or sold on one of several futures exchanges. The contract reflects estimates of the price of specified goods in coming months. In addition to energy, futures are traded on many agricultural products, livestock, precious metals, and even noncommodity items such as foreign currency and stocks. The concept is that when you buy a contract, you expect to exercise it and take delivery of the commodity at the fixed price, but in practice, futures contract buyers rarely exercise. Most contracts are sold before delivery date. The same qualification applies on the sale side. If you sell a futures contract, you are accepting an obligation to deliver the commodity at the fixed price. But the majority of these contracts will be closed or rolled forward, and delivery will not occur.

The greatest volume of activity in futures is in price speculation. If you buy a contract, and the price of the underlying commodity rises, you will be able to sell at a profit. If the price falls, your contract loses money. If you sell a futures contract, you will realize a profit if the underlying price falls. Because a seller enters the trade by first selling, the seller expects the price to fall so the contract can be bought to close at a lower price; the difference between the initial sale and the final "buy to close" will be profit. If that price rises, the seller has to close at a loss or roll the contract forward to a later date.

The contract contains several important terms. Among these is the specific commodity (or the currency or index on which the futures contract is written). The commodity in the case of energy futures is one of the several types of energy traded by way of futures (and there are many). Another term is the delivery date, the date on which the commodity is set for delivery or the contract must be closed. This date is known in the options market as expiration date. The contract also sets down the price per contract and current price. At the time of entering a contract, the futures contract's value is based on (a) the time to delivery date and (b) the span between the fixed price in the contract and the current value of the commodity. Listings of futures for each type of energy will show several different delivery dates, and the farther away that date, the higher the price. Time has value in the futures market, just as it does in the options market. So in order to fix a price way ahead, that time value is higher for longer-term contracts, and it declines as the delivery date approaches.

The futures contract is not exclusively speculative, although speculation does play a key role in the market. It is a matter of controversy whether speculation causes commodity prices to rise, as some politicians and television personalities have claimed (none of whom have any experience in the commodities market). However, it is more reasonable to observe that speculation is a reaction to the forces of supply and demand for commodities and is based on market fundamentals. In the case of energy futures, speculative trading is further driven by geopolitical developments, domestic politics, environmental interests, and other fundamentals not directly tied to actual demand. So there are two distinct actors in this market. The minority are organizations that want to fix commodity prices in the future and that intend to take delivery or to sell the commodity itself. But the larger group of actors are speculators, those who buy and sell contracts with the intention of closing before delivery date at a profit, accepting a loss, or rolling forward (closing the soon-to-expire contract and replacing it with a later-expiring one).

Trading is not limited to futures contracts, which can involve very large sums of cash. Trading also occurs by way of options on futures, in commodity index funds, and exchange-traded funds. Within these and notably in the ETF market, specific ETFs (exchange-traded funds), ETNs (exchange-traded notes), or ETCs (exchange-traded commodities) are designed to focus on a very narrow range of energy futures. These include baskets of securities in one type of futures contract, those with more than one energy type (multicontract ETFs), bearish ETFs (those whose value increases if the basket of securities falls), and very focused ETFs (such as those for clean and alternative energy), and some broader ones (for example, some include positions to track the entire energy sector). These are explained in greater detail in chapter 7.

THE DETAILS OF A FUTURES CONTRACT

Energy futures, whether bought directly or through one of the pooled fund alternatives, are risk transfer devices. The contract anticipates prices several months ahead and allows you to take a position at a fixed price in the knowledge that the price is going to change (either for you or against you). Knowing this affects your decision to buy or sell a futures contract or a broader position in a basket of futures. Very few individuals are likely to become involved in direct trading of futures contracts as this can carry great risk and be very expensive.

Tip: Most investors and traders will not want to trade futures contracts directly due to the high cost of many contracts. Alternatives are easier, cheaper, and better diversified.

Futures positions, especially through ETFs or commodity funds, can also be used to set up a hedge against your portfolio. For example, if you hold positions in energy stocks in your portfolio, selling a futures contract (or buying shares in a bear ETF) hedges against potential market risk in the stock position. If the stock values decline, the bearish position in futures will increase and thus offset that loss.

For the majority of traders and investors, these alternatives are better diversified, less risky, and less expensive than direct ownership of futures contracts. The level of trade required for single futures contracts can be quite high, a factor that by itself excludes many people from participating directly in this market. The notional value (full value of a single contract) can be quite high. For example, a single contract for light crude oil is based on 1,000 barrels. If the current price is $90 per barrel, the notional value is $90,000. However, the initial margin and maintenance margin requirements are much smaller.

This points out the high risk of margin leverage, however. A big move in the futures price accelerates the margin requirement. If the notional contract price is $90,000 but you can buy a single contract for margin under $10,000, is that an advantage? It is more likely to be an indication of much higher risks. Leverage is convenient, but it adds to overall risk. This is why a majority of futures traders will be attracted to commodity funds or ETFs as opposed to direct trading in futures contracts.

Another element of risk involves shorting futures. Buying a contract (going long) is certainly a risky venture. However, selling (going short) can be even more risky. In either buying or selling contracts, you can pick from a series of delivery months, with the closest one being the most sensitive to price changes.

Futures prices are supposed to change based on the perceived and actual values of each commodity. With improvements in technology for energy (and agriculture), past predictions of shortages and even famines have been proven wrong. For example, in 1967 a prediction was made that between 1970 and 1985, shortages of all necessities would result from population growth. This effect would be seen in energy and agriculture prices. However, improved technology in oil extraction and improved agricultural technology have demonstrated that predictions are not only difficult, but often entirely wrong. The futurist making those predictions published his ideas, and this led to the zero population growth movement. Today, declining birth rates, not a population boom, are seen as the problem.[3]

The inaccuracy of this prediction on a long-term scale demonstrates how difficult it is to predict prices even a few months ahead. The interesting aspect of futures prices is that they are set in an attempt to bring order to the actual delivery of commodities in the market, and many companies that rely heavily on energy for transportation rely on futures contracts to fix today's price for future purchase.

WHAT IS THE FUTURES MARKET?

The energy futures market is the largest segment in the commodities business. Over 100 million oil and gas futures contracts are traded each year, plus 25 million in natural gas. In comparison, corn trades total 47 million, soybeans 26 million, and sugar 18 million. The futures market is a high-volume market and includes not only direct buying and selling but also a very high volume in options on futures.

The seven most active energy futures accounted for over 563 million trades in 2011, and there were many additional energy futures as well. These seven are summarized in figure 5.2.

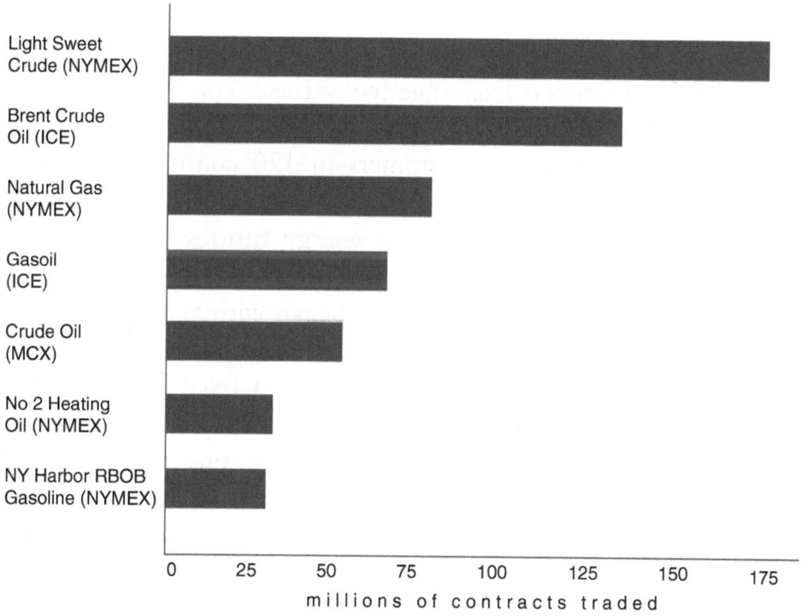

Figure 5.2 Seven largest-volume energy futures

Source: prepared by author from 2011 raw data at Will Acworth, "Annual Volume Survey – Volume Climbs 11.4% to 25 Billion Contracts Worldwide," Futures Industry, March, 2012

Today's futures market is truly global. Before 2000 many exchanges offered futures trading regionally, but no global markets existed. This has changed in only a few years. In 2011, as many as 12.9 trillion futures contracts traded worldwide, and of these a total of 814.77 billion were energy futures.[4]

Today, many of the previously separate exchanges have consolidated. The largest physical futures exchange based in the United States today is the CME Group (www.cmegroup.com), a holding company for five contract markets: Chicago Mercantile Exchange (CME), New York Mercantile Exchange (NYMEX), Chicago Board of Trade (CBOT), Commodity Exchange (COMEX), and Kansas City Board of Trade (KCBT).

Tip: The trend toward consolidation of futures exchanges is a positive one for investors. By combining trading in fewer places, prices are more uniform, and ultimately the price of trading may be reduced as well.

In addition, a lot of commodities trading takes place on a global scale. The International Exchange (www.theice.com) operates over-the-counter markets in North America and Europe, with trading firms in 70 countries and customers in 120 countries. In 2007, ICE purchased and merged with the New York Board of Trade (NYBOT). More than half of the energy futures worldwide are traded on ICE.[5]

Futures trades in the United States began with agricultural markets in the nineteenth century. Chicago became the center for futures trading, where farmers could be assured of a fixed price for their crop yields in the coming season. At that time, the United States was primarily an agricultural society. In 1790, the census revealed that 95% of the population lived on farms, and most of the 5% "urban" population lived in small villages. Only Philadelphia, New York, and Boston had population totals above 15,000. By 1890, the urban population had grown to 35%, but most citizens had never traveled more than 20 miles from their birthplace. But by 1990, 75% of citizens in the United States lived in urban centers.[6]

This growth pattern reveals how the United States evolved from an agricultural society into an industrialized, urban society. An increased reliance on energy has been a part of the change, brought about largely by developments such as the automobile and the airline industry, and other twentieth-century changes in US society. This explains why the Midwest agricultural commodities markets have gradually evolved into a commodities market dominated by a combination of agriculture and energy.

The futures market has played a key role in creating an orderly market, which today involves not only buyers and sellers of futures with delivery in mind but also a high-volume market in speculation. The market itself has evolved in other ways, too. In the nineteenth century, farmers and ranchers entered into *forward* contracts, informal agreements between buyer and seller laying out the price of the goods but not with specific deadlines or amounts of agricultural product to be traded. These were entered into individually, so there was no liquid market for forward contracts. This explains why, as the economy has moved more toward energy (along with agriculture), the more formal futures contract has become the norm. It offers standardized terms: delivery/settlement date, initial contract price, and amount of goods per contract.

The futures market has changed rapidly since the middle of the twentieth century. In the 1970s, financial futures were introduced, allowing traders to speculate on interest and currency rates. The expansion of the Internet has enabled this high-volume activity to expand rapidly, so that all futures trading is now quick as well as efficient. No longer do traders rely on brokers to represent them, and open outcry (the practice of vocal bidding for prices between brokers) is quickly being replaced by fully automated bidding systems.

In the nineteenth century, the largely agrarian society of the United States defined the commodities market. The expression, "cotton is king" described the economic base of US commerce, both at home and abroad. Today, however, although agriculture continues to play an essential role in the economy of the United States, energy has taken at least an equal role in the economy, often defining how expansion occurs and at what rate.

Key point: Just as cotton dominated the nineteenth-century economy, energy products are likely to dominate in the twenty-first century.

Another difference between the nineteenth and twenty-first centuries is found in methods of moving goods to market. Before autos and trucks became mainstream, trains provided a primary method for transporting goods. Steam and coal were the primary sources of power. In 1825, the Erie Canal, a 363-mile connector between Lake Erie and the Hudson River, connected the Atlantic Ocean with the Great Lakes, making Chicago accessible to ocean-bound importing and exporting. This was the most significant transportation improvement in US history, and its success ushered in the "age of the canal." The Erie Canal was responsible for more than movement of goods. Immigrants also used it to migrate west and the great westward expansion resulted. Chicago and St. Louis became gateway cities between the Eastern United States and the territories in the West.

These important advances in transportation systems affected movement of goods as well as of the population and led to today's heavy reliance on energy futures. Although people now move about with more efficient modes of transportation, the cultural reliance on mobility defined the role of the energy market in the nineteenth century and still does so today. The Erie Canal also changed the entire futures business. With cheaper and faster routes from farm to market, agriculture was revolutionized by the canal.

Even with big advancements in agricultural futures made possible by canals and railroads, the futures market remained fairly limited. The time required for movement of perishable goods kept agriculture limited. The use of grain storage silos helped preserve some products but not all. It was only when the automobile age and the aviation age began that agriculture became truly efficient, and that change depended on the use of energy products.

Risks and Rewards of Futures Trading

Even with the many advances in technology and transportation, the futures business has not always kept pace with social change. Direct transaction of energy futures is expensive and carries high risk. The contract size for many futures is prohibitive for most traders. For example, crude oil futures trade at 1,000 barrels per contract; heating oil is 42,000 gallons, and natural gas trades at 10,000 BTUs per contract. Even with margin, the level of cash required is quite high. Secondly, the trading costs of futures make it very difficult for individuals to earn a profit by direct trading.

The rewards of futures trading come in the form of potentially fast profits. When prices move quickly—as they do in the energy market—traders can make fast profits from the leverage in futures contracts, and this is possible whether the market price moves up or down.

Futures contracts can be used for two purposes. For the provider of energy and for the consumer, the fixing of a future price brings order to the market. Originally designed to fix the price of agricultural products even before crops were planted, the concept has become popular in many industries, and today fixing prices to make them predictable is a cornerstone of the futures market. The second purpose is to provide speculators with a market for bullish and bearish positions. This is accomplished by individuals through trading in options on futures, exchange-traded funds, or commodities index funds. Very few individuals could afford the high contract price and margin requirements of energy futures.

Tip: Just because energy futures prices are high on a per-contract basis should not leave anyone out of the market. There are many practical alternatives to buying or selling futures directly.

The risk and return of energy futures has to be quantified in terms of increments as well as percentages. If you are required to place thousands of dollars of margin against a $90,000 increment for a single futures contract on crude oil, the risk is substantial even if you believe the price is going to move favorably. This is true for two reasons:

1. Leverage is risky. A $90,000 contract value can be opened for only a few thousand dollars, but therein lies the problem. Each point that the future value moves against you is multiplied when you leverage the position. This adds to the risk. Traders may fall into the trap of seeing high levels of leverage as a great advantage, because they consider the profit potential, but they

tend to overlook the risk exposure. With margin as low as 5% of the contract value, risks are very high.

2. At this level of capital, it is difficult to diversify the exposure. Diversification is always desirable and, in fact, is built in to both ETF and commodity fund products. In comparison, direct transactions in futures contracts are very difficult to diversify effectively due to the high cost and high leveraged risk exposure.

PROFITABILITY TESTING

The big challenge for traders in the futures market is deciding in which direction the market is going to move, and judging profitability of trades. If market prices rise or fall significantly (as they often do), traders need to decide whether to take a bullish or bearish position or whether to set up a combination of positions that might be profitable regardless of price direction. These "hedges" involve long and short option contracts or long and short positions in energy-related exchange-traded funds.

Any evaluation of profitability should be made based on comparisons between futures and other products (stocks or bonds, for example). An evaluation of stock profits incorporates capital gain/loss as well as applicable dividends. Bond returns include interest as well as the cost of premium or income of discount. However, profitability calculations also have to take into account the cost of the transaction. Bond costs vary depending on where you buy and sell. For stocks, if you use online discount brokers, the transaction is going to average about $10 to buy and another $10 to sell. Compared to direct trading of futures, stock trading is quite economical.

The comparison of trading costs may also be made between direct trading in futures versus trading in an exchange-traded fund (with better diversification in energy) or commodity index funds (with diversification among a broader set of commodities). You can trade these at about the same cost as shares of stock. In fact, if you prefer the debt market, you can combine the low trading costs of ETFs with the debt position by focusing on an ETN (exchange-traded note) rather than the equity-based ETF.

With today's low interest yields, the comparison between equity and debt might not even be valid. Historically, treasuries have served as a comparative benchmark. But today, this is no longer practical, given how low these yields have become. There are other means of comparison, however. For example, many ETFs offer options on the ETF basket of securities.

Energy futures are thought to carry high risk, based on trading methods of the past (direct buying and selling of contracts). Today, with so many alternative products (ETFs, commodity funds, and options, for example) commodity traders have many other choices. You can even trade in commodities by investing directly in shares of energy companies. The next chapter explores this alternative.

CHAPTER 6

ENERGY STOCKS

> There will be a major positive shift in the world energy equation whereby technological breakthroughs of fracking and horizontal drilling have opened up vast new potential supplies of natural gas and oil around the world. Dependence on the volatile Middle East will be reduced, and the perceived need for unproductive "green" energy investments will be reduced. Energy will not be a constraint on global economic growth longer term.
>
> Peter T. Treadway and Michael C. S. Wong, *Investing in the Age of Sovereign Defaults* (Wiley) 2013

THE ENERGY MARKET CAN BE ENTERED IN MANY DIFFERENT WAYS, some are high-risk investments (direct trading of futures, for example), and others have built-in diversification (such as ETFs or index funds). In the middle of this risk/cost range is the traditional direct ownership of stock.

Buying stock may present one of the best opportunities for profit, but it also carries many specific risks. If stockholders know these risks and how to manage them, they can exert great control, time their market entry and exit, or adopt a long-term buy-and-hold approach to investing. Direct ownership of stock is quite flexible and liquid, and being able to trade quickly is one of the big advantages of the stock market.

Key point: Opportunity and risk are two sides of the same coin. The higher the opportunity for profit, the greater the risk of loss.

The best way to reduce risk is to develop a sensible program for critical evaluation of companies and their stock. Two methods used are fundamental and technical analysis. These are based on a study of financial trends and financial reports and on price patterns and movements of stock.

What Are Energy Stocks?

Energy stocks, like other stocks, represent part ownership in the corporation. Each share is an extremely small piece of the whole.

> **Tip:** What percentage of the company is a single share of stock? A share of Exxon Mobil (XOM), the biggest integrated oil company, was worth about $88 at the beginning of May 2013. With Exxon's $393.4 billion of total capitalization (market value), each share is a very small percentage of the total: 0.0000000002237%.

Owning stock is a way for individuals with very little money to take part in the free market system. Stock is traded on the exchange, and if you own one share of Exxon or of any other energy company, you are an owner of the company, although your share is only small.

As an owner of stock, your tangible share gives you a piece of the total, but that doesn't mean you can claim any of the company's assets. You can't walk into Exxon and demand a desk or office supplies in exchange for your $84 share. Your share is part of the entire deal, not identified with any specific assets. This is the nature of stock investing. You own a small piece of the corporation, but not any specific corporate assets. Corporate shares enable large companies to fund their operations on a massive scale while solving the problem of capitalization. Where would Exxon get $393.4 billion if the company had to rely on only a handful of investors? The public markets facilitate trading with publicly traded stock in large corporations. This means that the exchanges (public markets) match up buyers and sellers to ensure that both sides are able to trade, and to negotiate prices. Owning stock also comes with a few benefits. If the value of Exxon stock goes up, your share price also rises, and your investment is profitable. You're entitled to dividends. As of 2013,

Exxon pays 63 cents per share per quarter or $2.52 per year. That's not a big cash bonanza, but it's cash you can take out, or you can reinvest the dividend in additional shares. That quarterly dividend will buy you about three-quarters of an additional share four times a year. For example, as of September 2013 Exxon was trading at about $87.50 per share. A quarterly dividend on 100 shares of $63 would buy nearly three-quarters of one share ($63 ÷ $87.50 = 0.72). That is, by reinvesting dividends in the purchase of additional partial shares, you increase value. Over one full year at the same price levels, your 100 shares grow to 102.88 shares:

1st quarter: 100 shares + 0.63 shares = 100.63 shares
2nd quarter: 100.63 + 0.6340 (100.63 x 0.63) = 101.26 shares
3rd quarter: 101.26 + 0.6379 (101.26 x 0.63) = 101.90 shares
4th quarter: 101.90 + 0.6420 (101.90 x 0.63) = 102.54 shares

By reinvesting dividends to buy additional partial shares, 100 shares grow to 102.54 shares.

Stockholders also get to vote in matters the board of directors decides at its quarterly meetings. That small percentage of ownership is not going to affect Exxon's operations by itself, but you have that voting right, and that is considered a big deal by investors.

This voting right, like the dividend, is among the benefits of owning common stock. A risk is that if the company were to fall apart and go broke, *preferred* stockholders get paid off first, then bondholders (who loan money to the company), and finally common stock investors.

Chances of Exxon going broke in the near future are pretty slim. The $393.4 billion in market cap is huge, and the company is larger than its competitors. For example, Royal Dutch Shell has market cap of $124 billion, and Chevron is at $233 billion, both quite small compared to Exxon.

By the end of 2012, Exxon was the most profitable company in the Fortune 500, Wal-Mart ranked second, and ConocoPhillips ranked third—that is, two of the top three earners were energy companies.[1]

Combining its exploration and production (64% of activity), refining and marketing (28%), and chemicals (8%), Exxon reported 2012 revenues of $482.3 billion and earnings of $44.9 billion.[2]

These are impressive numbers, and as you might expect they create a buzz in the market and translate to a lot of buying and selling.

However, many different individuals and institutions are involved in trading shares of stock in a multitude of energy companies.

The energy sector consists of the following major subsectors:

Integrated oil and gas: This subsector is involved in all phases of oil and gas production, including exploration, production, refining, and distribution. Major players include ExxonMobil, Royal Dutch Shell, Chevron, ConocoPhillips, Petrochina, and Petroleo Brasileiro.

Oil field services: This subsector offers support services to other oil and gas companies, including extraction, drilling, and pumping. Some of the major companies in this group are Halliburton, Paramount Petroleum, Schlumberger, and Superior Energy.

Refining: No oil gets to the market until it is refined. Crude oil is broken down into many different types of fuel, and the refining process determines how much product gets to the market. Major refinery companies include BP, Chevron, Citco, ExxonMobil Refining, Houston Refining, Murphy Oil, Sunoco, and Valera Energy.

Mining: This subsector covers all energy products extracted from the earth, such as coal and uranium. Among the major mining companies are Alliance Resource Partners, American Electric Power, Black Hills, General Dynamics, Massey Energy, and Peabody Energy.

Renewable energy: This subsector covers a range of specialized energy types, including solar, wind, and hydro. Among the companies involved are ArcherDaniels Midland, Babcock & Brown, Ballard Power, First Solar, and SunPower.

Utilities: This subsector provides for the delivery of energy in the form of electricity; utility companies are major energy sector players. There are many of them. Among the biggest are American Electric Power, Atlantic Energy, Boston Edison, Con Edison, Duke Energy, Gulf Power, Illinois Power, Ohio Edison, and Virginia Power.

WHO IS INVOLVED IN BUYING AND SELLING?

When shares of stock are traded through the exchange, the increments of a single share of stock are very small, and trades go in bigger blocks from 100 shares traded by individuals up to 10,000 shares or more traded by institutional investors (such as mutual funds, administrators of pension and profit-sharing plans, and insurance companies).

The stock market is called an "open auction" market because prices change continually based on activity among two groups: buyers and sellers. Basically, there are two types of investors, institutional and retail. Institutions trade most of the dollar value of the market, but high-frequency trader (HFTs) investors (individuals) execute more trading activity. Retail (individuals) account for only 11% of trading volume. HFTs do most trading at 56% of the market; hedge funds account for 15%, and institutional trading represents 18% of volume.[3]

High-frequency trading consists of applying automated tools and algorithms to trade at lightning speed. Positions often are open only a few seconds or even fractions of a second, all these HFT systems need, and the volume of trades can run into thousands per day. It is the ultimate day trading strategy, since no positions remain open by the close of each day's session. The strategy is based on automated price tracking and high liquidity through the arbitrage system.[4]

The HFT trader needs only a small fraction of one cent per share to create rapid profits based on fast trading and a high increment level, executed many times per day. The process is based on daily price changes as well as on systems tracking prices over a longer term of weeks or months. This is the "algo" that identifies those split second opportunities based on purely technical price movement and changes.[5] Considering the rapid in-and-out moves of HFT traders, it would seem that fractional exploitation of price movement would not do much harm to retail investors who do not have access to the same technology. However, some critics have identified HFT as the cause of new volatility in the markets, which clearly may present greater market risk to small traders.[6] The widely publicized "flash crash" of May 6, 2010, was blamed in part on HFT activity, specifically the HFT rapid withdrawal from the markets right before the crash occurred.[7]

The opinion that HFT affects stock prices adversely has been recognized internationally and not only in the United States. In July 2001, a report issued by the International Organization of Securities Commissions (IOSCO) stated that "algorithms and HFT technology have been used by market participants to manage their trading and risk, [but] their usage was also clearly a contributing factor in the flash crash event of May 6, 2010."[8] This conclusion was further supported by the Chicago Federal Reserve, which stated that "every exchange interviewed had experienced one or more errant algorithms."[9] The Fed also recommended limiting the number of orders that could be placed within a specified period of time to avoid this problem in the future.[10] Many European exchanges have also recommended curtailing HFT trading volume because of its role in creating volatility.[11]

HTF is only one of many challenges stock traders face, and in the high-volume energy market the potential risks may be even greater than in the market as a whole. In the absence of volatility, market price movement is determined by the relatively simple forces of supply and demand. These economic forces create a daily struggle between buyers and sellers, and this creates price movement. For individuals investing relatively modest sums of money, the reality is that the big institutions with their 10,000-block trades determine price movement for the most part, with the individual following the trend. Even so, as individual investor you have a great advantage: You can move your money around instantly, buying and selling shares, usually with orders getting placed within seconds from the keystroke. Institutions have a harder time when moving thousands of shares. Those big blocks are not always going to get the price the institutions want when they sell, because a large block on the market tends to make the price fall. By the same token, when an institution wants to buy 10,000 shares, that higher demand tends to drive up the price. Accordingly, institutional trades are likely to have an immediate effect on price, not only for the entire market, but also on the institutions' daily trades.

The third party in all of this is the exchange and brokerage structure. Traders place trades with brokers and pay a trading fee, and brokers then execute trades through specialists in the stock exchange. It all happens very quickly in most instances. Online trades usually are finalized within seconds after an order is placed.

Individuals may place trades based on their own analysis and decisions, or they may work with a broker. The amount of market information you have before you make a trade can be modest compared to what institutional investors have. Large companies have not only an expert executive managing the portfolio, but they often have an entire department of researchers and market observers, all working to get the best prices and examine the details for many companies whose shares the institution will buy or sell.

As a small individual investor (or, as the market refers to you, a "retail" investor) you need to decide how to pick stocks you will buy and hold. Among the criteria investors rely on are dividend yield, profitability, sector, and position in the sector (for example, Exxon dominates the big oil sector). Should you always buy the biggest or most profitable company? Or are you better off focusing on second tier companies whose profits might grow more rapidly? These are questions every investor and trader has to ask in the process of buying stocks.

Buying Stocks Directly

Buying shares in a publicly traded company brings great opportunity for profit and also great risk of loss. With this in mind, you need to develop a system for comparing stocks of companies and evaluating their risks. For most investors and traders, this means studying fundamental and technical trends.

A comparison between three oil companies shows some of the immediately recognizable differences. Exxon Mobil (XOM), Chevron (CVX), and ConocoPhillips (COP) are all in the same industry, but their attributes are quite different. For example, as of the end of April 2013, dividend yield (dividend declared per share, divided by price per share) for each of these was shown in Table 6.1.

These dividend yields might seem fairly close together. But the dividend on COP is 57% higher than the dividend on Exxon Mobil. This is a substantial difference; dividend yield is a major factor to consider when picking one stock over another. As a result, if all other comparisons are equal, dividend yield may be used to select

Table 6.1 Dividend yields

Exxon Mobil	2.86%
Chevron	3.33%
ConocoPhillips	4.48%

Source: S&P Stock Reports

a company as a way to invest in energy. Comparing dividend yields of any of the three major oil companies to yields on other financial products, such as savings accounts, certificates of deposit, or Treasury securities shows that these yields are favorable.

Differences in dividend yield also have to be considered when deciding between energy stocks and stocks in other sectors. This is more difficult, because many companies have high yields. Many non-energy sector stocks yield dividends between 4% and 6% and even higher; anyone who seeks current income in addition to sound growth prospects may find that diversifying among many sectors is a good strategy. Using dividend yield to make comparisons is also a wise method for stock picking.

However, caution is also essential. High yield can be deceptive, and when a company is losing its competitive edge and prices begin to fall, dividend yield will rise. As a result, what at first appears to be an attractive yield could reflect an overall poor investment. For example, if a company's stock is price at $80 per share and it pays $2 per year in dividend, that's a 2.5% return ($2 ÷ $80). However, if the company is failing and the stock price declines to $66, dividend yield then *rises* to 3.0% ($2 ÷ $66). This big increase in yield is a consequence of the underlying weakness in the company and its stock.

This means that when you pick one stock over another, it is dangerous to look at any one indicator. A high dividend yield might represent a strong and competitive current income, or it could be the result of a weakening fundamental position in the industry. You should conduct additional research before deciding to buy shares.

Tip: To calculate dividend yield, divide the annual dividend per share by the current price per share. If today's dividend is $2 and the current price is $80 per share, dividend yield is 2.5%: ($2 ÷ 80 = 2.5)

Comparisons between stocks are complicated by the fact that the industry is not made up of companies all operating in the same field. For example, in the energy sector there are many subsectors, and these face varying kinds of market risks, competition, and costs.

A detailed breakdown of the energy sector includes companies in many industries:

Major integrated oil and gas
Oil and gas pipelines
Oil and gas equipment and services
Oil and gas drilling and exploration
Independent oil and gas
Oil and gas refining and marketing
Natural gas
Coal mining
Coal distribution
Electricity generation
Electric power distribution
Renewable energy
Ethanol and biofuels
Alternative energy
Nuclear power

STOCK RISKS AND REWARDS

The risks of direct ownership of energy stocks depend on the current supply and demand as well as on the subcategory of energy company involved (diversified oil, refineries, oil service, and exploration, for example). As with all stocks, direct ownership should be based on comparisons of fundamental trends over many years.

You can make money very quickly by trading energy sector stocks. But this is true due to volatility in the market, which means that losses can take place just as quickly. Anyone trading in energy stocks needs to know the risks as well as the rewards.

Stock valuation is not just a matter of price or short-term price trends. A $30 stock is not twice as valuable as a $15 stock; the number you need to look at is market capitalization or "market cap." For example, a $30 stock with 80 million shares outstanding has a market cap of $2.4 billion. And a $15 stock with 160 million shares outstanding *also* has market cap of $2.4 billion:

$30 x 80 million shares = $2.4 billion
$15 x 160 million shares = $2.4 billion

Key point: Stock price all by itself is useful only for tracking changes in that one stock. You cannot tell anything from comparing one stock's price to another. You need to compare market cap.

Another way to judge a stock and the risks involved in buying it is on the basis of the price/earnings ratio (PE). This is a calculation in which the price per share is divided by the latest reported earnings per share, or EPS. For example, if a stock's current market price per share is $30 and the most recent reported earnings are $2.14 per share, P/E is 14:

$$\$30 \div \$2.14 = 14$$

The P/E is one of the best ways to decide whether a stock's price today is a bargain price or is too expensive. The multiple (in this example, 14) reveals that the price is equal to 14 years of earnings, based on the most recently reported earnings per share. As a general guideline, a P/E between 10 and 25 is moderate. Anything below 10 implies very little interest in the company; as the P/E moves above 25, the stock gets very expensive.

There are a few problems with PE, among them the following:

1. *The trend is where you get insight.* You cannot judge a stock based on today's PE. You need to study the *range* of low-to-high P/E ratio per year for several years to find the trend. The difference between high and low P/E shows you how volatile both price and earnings have been over many years.
2. *The timing of each side of the equation is different.* P/E as a single "value" is inaccurate because it compares today's price to earnings from several weeks, possibly months, earlier. Earnings may also be adjusted even after financial statements are published, making it even more crucial to study a long-term range rather than today's P/E alone. So P/E equates today's price with recent historical earnings even though a lot might have changed since the fixed date of those earnings reported most recently.
3. *By itself, P/E is not a reliable indicator of a stock's value.* P/E is only a yardstick. It is valuable as one of many indicators you

need to consider when choosing stocks. A stock whose P/E is very high is going to be overpriced, and this makes it easy to eliminate that company from a range of possibilities. But a stock with a midrange P/E is not always the best choice; you need to review other fundamentals as well.

In the energy sector, market risks and cyclical changes are neither singular nor universal. Many different market and economic attributes affect the subsectors of energy in different ways. The best way to judge a subsector's health is by comparing several companies that compete with one another directly; then you can get a sense of what is typical and look for companies that dominate and outperform others. These are likely to be your best bets.

Key point: The energy sector is not a single set of companies all operating in the same cyclical, economic, or political climate. The subsectors often are quite different from each other and are valued based on separate forces.

Some investors, instead of picking one or two companies, prefer buying energy exchange-traded funds (ETFs). These are mutual funds that specialize within a single market sector (or region or type of security), and you can find an ETF with a highly focused and specialized basket of securities, the term used for a portfolio in the ETF market because it is defined in advance. In comparison, traditional mutual funds are managed, and the portfolio may change often. An energy ETF solves the problem of selection and builds in diversification. On the downside, if you buy shares of an ETF, your earnings are going to be the average of that basket of securities, including the average of overperformers and underperformers.

If you are determined to own energy stocks directly, the best approach is to identify a few important fundamental indicators and track stocks over several years to spot trends. If you can track a stock for ten years, you get a good idea of whether financial matters are improving or declining. Services like the *S&P Stock Report* provide a ten-year summary of all of the key indicators. This is provided free to members of most of the online discount brokerage services.

Tip: One of the best ways to track a company's financial trends is to study the ten-year history provided by Standard and Poor's in their *S&P Stock Reports.* These are provided free of charge to anyone with a trading account at Charles Schwab, TD Waterhouse, Scottrade, and e*Trade, for example.

Is It Profitable? How to Identify Stocks That Work for You

Everyone who wants to buy and sell stock needs to decide what companies to buy, a seemingly obvious point but possibly the most elusive aspect of stock selection. In the energy sector, Exxon has been the most profitable company not only in that sector, but also in the entire Fortune 500. So if net profit is the only criteria, integrated oil and gas—especially the giants—are the obvious choices.

However, investors have to narrow down their choices even more. Hundreds, even thousands of companies are profitable, but how does that overall profitability affect stock prices? This is the big question. If you buy shares in a company hoping the price will rise and create portfolio profits, you probably begin with profitable companies.

However, this does not guarantee that share price will rise. The market does not respond only to financial success but even more to the supply and demand among investors and traders and to the *expectation* that future growth will be better than average. Even profitable companies can be perceived as having little or no growth potential, and in that case money will flow out of those shares and into shares of another company.

In other words, a company's profitability does not always translate into investment profitability. If it did, picking stocks would be easy.

The energy sector has a lot going for it. In terms of market perceptions, energy is among the best sectors for investors. Its product, the commodity of energy in its many forms, is a necessity in every home, for all transportation, and in all industries. The supply is limited, so prices rise as use rises. This is the fundamental reality of supply and demand, and it directly affects the prices of energy stocks.

Before you as an investor moves all of your investment capital into energy stocks, a few very basic steps should be taken. First, you need to identify your level of risk tolerance. How much risk can you afford to take? Many people view owning stock directly in any company as too risky because they fear that any one company's stock can lose value. For these investors, mutual funds and other "pools" make sense because diversification is built into the portfolio by the mutual fund's management. For others, the entire stock market (including energy stocks and mutual funds) is too volatile. These investors will tend to prefer buying gold coins, real estate, or agricultural futures.

There are many markets to pick from, and energy is only one. But it is an important one because its product is a basic necessity, and the volume of trades in the energy market are likely to be high, quick, and vibrant – in other words, volatile. For many investors, being a player in this market can be as exciting as making profits, or even more so. Therein lies the danger. It is easy to get a "casino mentality" in investing and trading and to become attracted to the potential more even than to actual profits.

Moderate or conservative investors usually focus on the fundamentals, those financial indicators and trends that summarize how a company is doing in terms of profits over time.

ATTRIBUTES OF STOCKS: FUNDAMENTAL AND TECHNICAL

The challenge in fundamental analysis is figuring out what to track and what to overlook. You probably don't want to study accounting for four years just to become an expert in reading balance sheets. And you don't have to. Services such as Standard & Poor's track important indicators so you can get the results of as many of these as you want without having to become an expert. Relying on a short list of dependable indicators is a sensible fundamental approach to stock picking.

> **Key point:** You are not going to get better information from tracking more indicators. The best approach is to identify a few revealing trends and focus on these as you collect data. Fundamentals serve you best for comparisons between companies, but none will provide you with a "sure thing" for profits. However, when indicators go negative, they tell you clearly what stocks to avoid.

Every investor selects his or her own short list of fundamental indicators to watch. The two major concerns in fundamental analysis are profitability and working capital. Successful companies should grow, both in terms of revenue and net profits. At the same time, a well-managed company should manage its cash flow to rely on a steady or declining level of debt. With these guidelines in mind, here is a list of four suggested fundamental indicators worth tracking:

1. Revenue and profits. All companies have to grow and remain competitive. So you should expect to see a long-term trend of increasing revenues and profits. In addition to the dollar amount of profits, the net return should remain consistent or grow, too. Net return is calculated by dividing profit by revenue. For example, if last year's net profit was $44,880 (million) and revenue was $482,295 (million), net return was:

$44,880 ÷ $482,295 = 9.3%

In fact, this was the outcome for Exxon-Mobil (XOM) in fiscal year 2012. But by itself this doesn't reveal the long-term trend for the company. The ten-year record of revenue, net profit, and net return is summarized in figure 6.1.

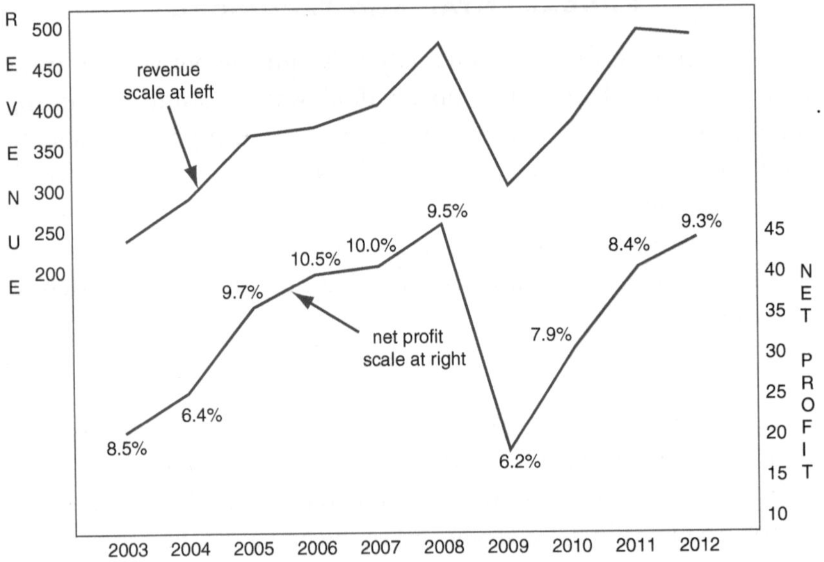

Figure 6.1 Exxon Mobil 10-year results (all dollar values in billions of dollars)
Source: Chart prepared by author from raw data at S&P Stock Reports

Just comparing the revenue and profit growth for the company shows that it was a very positive decade. The dip in 2009 was a market-wide decline, but otherwise the lines were moving in a positive direction. The net return was also positive, remaining in a desirable narrow zone.

Another set of results shows a somewhat different outcome but makes the point that the dollar levels are not as important as the overall trend. Chevron (CVX) had a ten-year record slightly different from Exxon Mobil's, as is summarized in figure 6.2.

The ten-year outcome of revenue, net profits, and net return for the third sample energy company, ConocoPhillips, is shown in figure 6.3.

A comparison of the three companies reveals how valuable a ten-year analysis is in selecting one company over another. Exxon was clearly the leader in terms of dollar value in both revenue and net profits. However, the net return of Exxon and Chevron was very similar throughout the period. In fact, the two companies had remarkably similar curves of growth in revenues. However, Chevron's net profits—although a smaller dollar amount—revealed stronger growth through the decade. ConocoPhillips' record was different from that

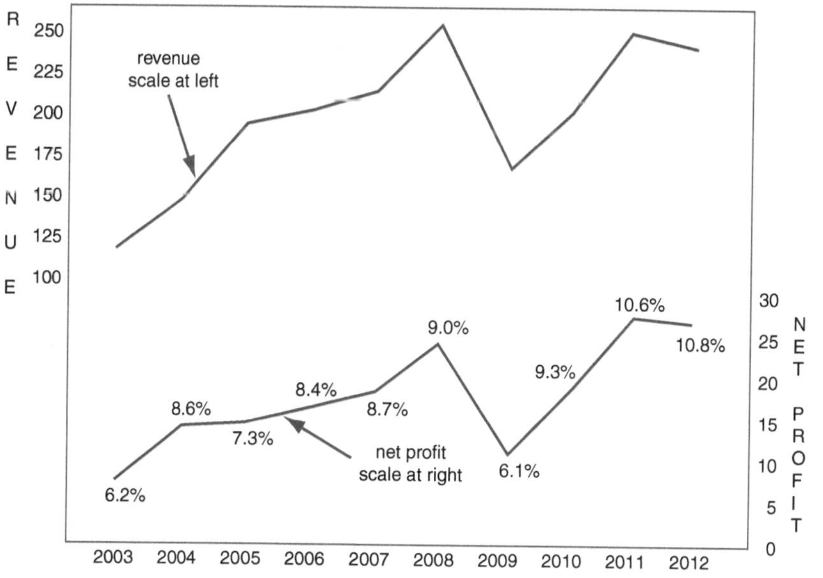

Figure 6.2 Chevron 10-year results (all dollar values in billions of dollars)
Source: Chart prepared by author from raw data at S&P Stock Reports

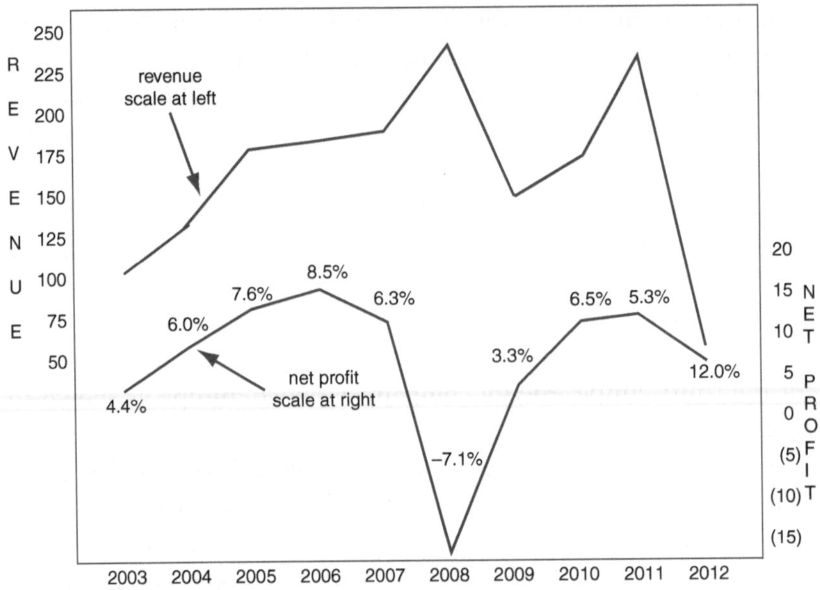

Figure 6.3 ConocoPhillips 10-year results (all dollar values in billions of dollars)
Source: Chart prepared by author from raw data at S&P Stock Reports

of the other two companies, not only in dollar values of revenue and net profits but also regarding net return.

COP pays the largest dividend of all three companies, but there is more to consider than dividend yield. The net profit record showed greater volatility, which translates to higher uncertainty and market risk. Twice during the decade, in 2008 and 2012, COP's revenues and net profits dipped substantially below the trend. Although all oil and gas companies reported similar declines, COP showed a more drastic fall. This is due in part to the smaller revenue and profit base, but it still translates to greater overall volatility as well.

In some respects, a larger market cap provides more power and lower volatility for Exxon, enabling it to expand and compete effectively with other companies in the sector. However, the larger a company becomes, the more difficult it is to make changes quickly. Thus, opinions vary on what size is best for a company or within a particular sector.

2. *Dividend yield* is a second important indicator to examine when choosing a company to invest in. The most desirable trend is an increase in dividends each year. That's a sign that management is able to keep working capital under control so that dividends can be paid to stockholders and can be increased over many years. A

comparison of Exxon Mobil, Chevron, and Conoco demonstrates how a ten-year history developed, as shown in table 6.2.

All three companies increased their dividend every year for ten consecutive years. This attribute is very desirable, but a more important test of dividends is the payout ratio. This is the result when dividends are divided by net income. Comparing the payout ratio for these three companies is more revealing than looking only at dividend per share, as is illustrated in figure 6.4.

Table 6.2 Dividends per share

Year	Exxon	Chevron	ConocoPhillips
2003	0.98	1.43	0.82
2004	1.06	1.53	0.90
2005	1.14	1.75	1.18
2006	1.28	2.01	1.44
2007	1.37	2.26	1.64
2008	1.55	2.53	1.88
2009	1.66	2.66	1.91
2010	1.74	2.84	2.15
2011	1.85	3.09	2.64
2012	2.18	3.51	2.64

Source: *S&P Stock Reports*, 10-year history through fiscal 2012.

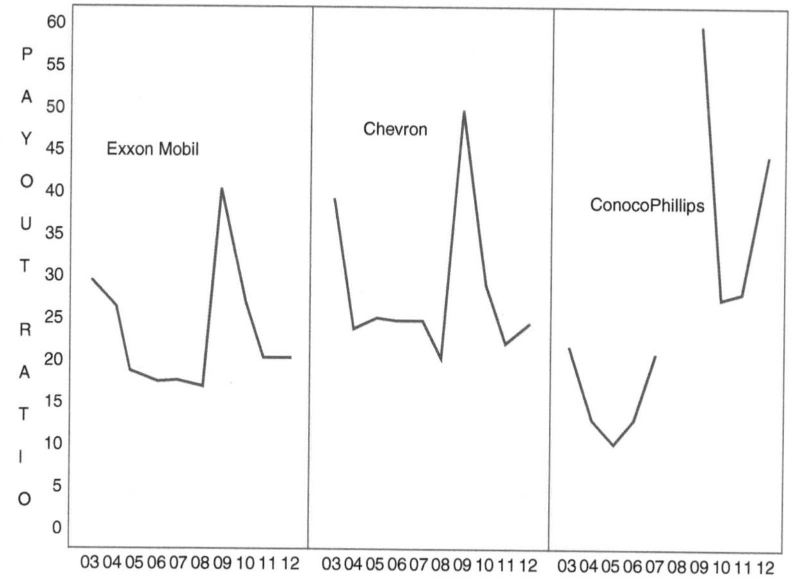

Figure 6.4 Dividend payout ratio comparison, 10 years
Source: Chart prepared by author from raw data at S&P Stock Reports

On this chart, you can see the spike in 2009 that all three companies experienced due to the sector-wide weakness and reduced revenues and profits. This had the effect of increasing the payout ratio, artificially moving the needle higher. In fact, the revenue and net profit trend was more revealing in this case than the payout ratio.

However, as a ten-year trend, the payout ratio reveals a comparison of greater significance than the dividend paid per share. For example, while Exxon's revenues and net profits doubled during the decade, the payout ratio declined from 31% in 2003 to only 22% in 2012. Chevron's revenues and net profits also grew, and the payout ratio fell similarly from 40% in 2003 to only 26% in 2012. ConocoPhillips reported more mixed results; revenues increased in general except for 2012, when they fell significantly. Net profits were inconsistent, and the trend was difficult to read. However, the payout ratio remained in the same range (when you remove the revenue and net profit spikes of 2009 and 2012).

Payout ratio thus should be viewed with a few caveats in mind. When revenues and net profits fall, payout ratio will rise, but this is not a significant *change* as much as a reflection of the movement of dollar values and net return. However, if you remove those spikes in revenue and earnings (as experienced in 2009), you get a longer-term view of what is going on. Although Exxon Mobil remains the largest of these companies and the most profitable, its payout ratio declined by about one-third (31% to 22%).

A third way to judge dividends is by consistency. A company that increases its dividend each year for at least the past ten years is considered a good value for investment and given the name dividend achiever. All three of these energy companies meet this definition.

> **Tip:** Companies that increase dividends every year are termed "dividend achievers," and are viewed as the best long-term investments. The concept was developed by Mergent Company (www.mergent.com).

Dividend achiever status reveals that the company manages its cash flow well enough to fund increasing dividends. However, this should be considered together with the payout ratio. Over time, dividend achievers have tended to report continued growth in revenues and net profits, more consistently than average companies. That is, dividend achievers are usually better investment candidates than average companies.[12]

3. P/E ratio is yet another important fundamental indicator. P/E tests how stocks are valued by the market. When stocks are above a P/E of 25, they are expensive. The multiple represents the number of years' earnings in the current price. However, to track PE, you need to study several years and not only the year-end P/E but the yearly range from high to low.

A study of this record for Exxon Mobil, Chevron, and ConocoPhillips allows you to draw conclusions about all three companies in terms of how the stock is priced, and you can see whether the companies are bargains at those prices or are overpriced compared to other stocks. An annual summary of the high and low P/E makes this point, as shown in table 6.3.

The comparison is revealing here. Exxon's range of P/E was quite consistent during the decade. In more recent years, the low side of P/E remained at 10. However, the range is more important than the P/E high and low numbers. Exxon remained in a range of four points or less with only one exception: 2009.

All three companies reported very low distance between high and low, indicating that the price levels are reasonable. When you see a huge change from high to low each year, this indicates higher risk for that company's stock. A small range in the P/E is favorable. However, a minimum level of 10 would be better than the single

Table 6.3 P/E ranges, 10 years

Year	Exxon	Chevron	ConocoPhillips
2003	13–10	12–9	9–8
2004	13–10	9–7	8–6
2005	12–9	10–8	7–4
2006	12–9	10–7	8–6
2007	13–9	11–7	13–9
2008	11–7	9–5	–
2009	21–16	15 – 11	18–11
2010	12–9	10–7	9–6
2011	11–8	8–6	9–7
2012	10–8	9–7	13–9

Source: S&P Stock Reports.

digits. Even so, when the full range is taken in to account, these three energy companies appear to be priced fairly by the market.

4. Debt ratio is another key fundamental indicator. This tests how much a company relies on borrowed money versus how much it is funded by equity or ownership on the part of stockholders.

Total capitalization is the combination of shareholders' equity (stock and retained earnings) and long-term debt (bonds issued plus long-term notes). The most desirable relationship is for debt to be relatively low and to stay at the same level year after year or, better yet, to decline over a period of years. A danger signal is a rising debt ratio, a signal that the company is relying increasingly on borrowed money. To compute the debt ratio, divide long-term debt by total capitalization (long-term debt plus equity):

Long-term debt ÷ (long-term debt + equity) = debt ratio

Key point: If you are studying a company and thinking of investing, what does it mean if the debt ratio is over 100? It means debt is out of control, and the equity value of the company is negative. When you see a debt ratio growing over a decade and approaching high levels, stay away. It is never a good thing for debt to increase every year.

The debt ratio is expressed as a single number without percentage signs, even though it represents a percentage. The history of debt ratio for Exxon, Chevron, and ConocoPhillips shows how the trend developed for both, as illustrated in table 6.4

Exxon's history is impressive. The company maintained long-term debt below 7% of total capitalization for most of the period. When a company, especially a large one like Exxon, is able to keep a lid on long-term debt, this is always a positive signal. Chevron's record is equally impressive but for a different reason. At the beginning of the decade, long-term debt was in double digits, but in the most recent five years, it was reduced. However, ConocoPhillips had the highest debt ratio of the three. At the end of the decade, the ratio was higher than at its beginning although in between the ratio fell. The percentages are revealing. While Exxon and Chevron had

Table 6.4 Debt ratio, 10 years

Year	Exxon	Chevron	ConocoPhillips
2003	4.0	18.9	27.2
2004	3.8	15.7	21.0
2005	4.4	13.4	14.1
2006	4.7	8.2	21.6
2007	4.6	7.3	19.3
2008	4.9	5.8	30.6
2009	5.5	–	–
2010	7.0	9.4	27.4
2011	5.2	–	–
2012	4.2	8.0	32.0

Source: S&P Stock Reports.

debt ratio in single digits, ConocoPhillips reported much higher debt levels.

Key point: Revenue/profits, dividend yield, P/E ratio, and debt ratio are only starting points in picking stocks. But as starting points, if you see weakness or decline in any of these, you are wise to look elsewhere.

Together these four key indicators (revenue/profits, dividend yield, P/E ratio, and debt ratio) serve as a starting point for evaluating and comparing investments. If you begin with these and then compare additional indicators, you will be able to narrow down your list to only a few possibilities.

Among the other tests worth applying are price levels and volatility. These technical indicators help identify market risk levels. If the trading range seen on the chart is wide or growing, then risks are greater as well.

Compare the one-year price charts for all three companies, beginning with the chart for Exxon Mobil in figure 6.5.

This chart reveals that even with price levels moving, the range is normally within four to six points. In other words, Exxon Mobil is not a particularly volatile stock, and its breadth of trading (distance between resistance and support) is quite small. Resistance is

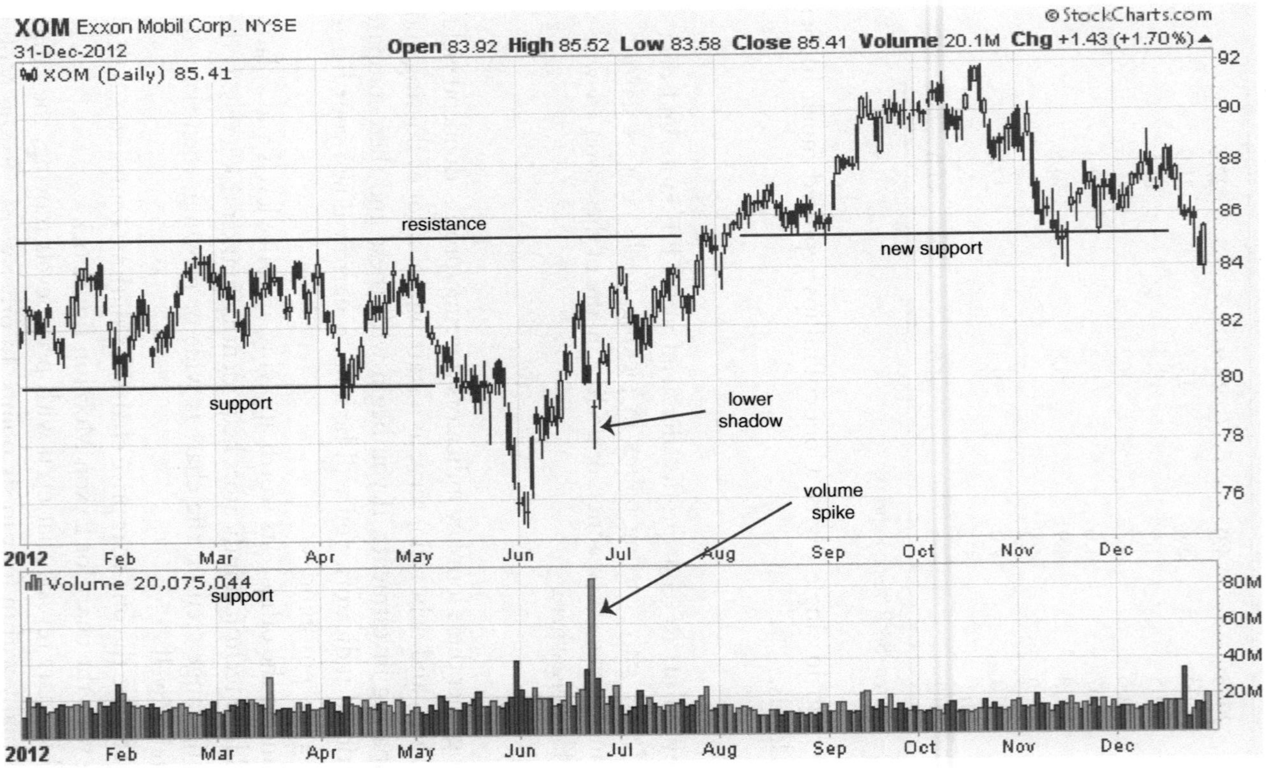

Figure 6.5 Exxon 1-year price chart

the highest price in the trading range, and support is the lowest. On this chart, notice that previous resistance became new support. This flip of the trading range is strong confirmation that the new bottom of the range (support) is likely to hold up.

A very strong reversal signal was found as well. The volume spike by itself is a strong indicator, but it was confirmed by the narrow-range day (opening and closing price were the same, also called a *doji* in candlestick analysis). The exceptionally long lower shadow (the range of trading below open or close) indicates lost momentum among sellers, who were not able to move price lower.

> **Tip:** The switch between resistance and support gives a technical testing level to decide whether a newly established trading range will hold or fall apart. If the new range holds up, it means the breakout was probably the real deal.

The chart for Chevron is also interesting. A series of relatively short downward-moving trend lines were offset by support levels rising through the trend. This is not an unusual pattern, but it is unusual to see it repeat so many times in a 12-month period. While this chart appears quite volatile, the range of trading was about six points, similar to what it was for XOM. Three of the important reversal points were marked by the end of the price trend and volume spikes occurring at about the same point on the chart. This chart is shown in figure 6.6.

The third company, ConocoPhillips, also had an interesting one-year chart, as shown in figure 6.7. Here you see a fast-moving trading range that never exceeded four points. A large price movement is bordered by two sets of volume spikes, both with sharp trend lines before a more gradual uptrend took over.

All three of these charts offer price trend insights with different type of indicators and with varying levels of momentum. In technical analysis, two general theories are used by chart analysts. These are Western (traditional signals such as tests of support and resistance, price gaps, volume spikes, and a pattern) and Eastern (candlestick patterns). These two approaches are most effective when used together. Neither approach by itself gives you everything you will need to anticipate price direction.

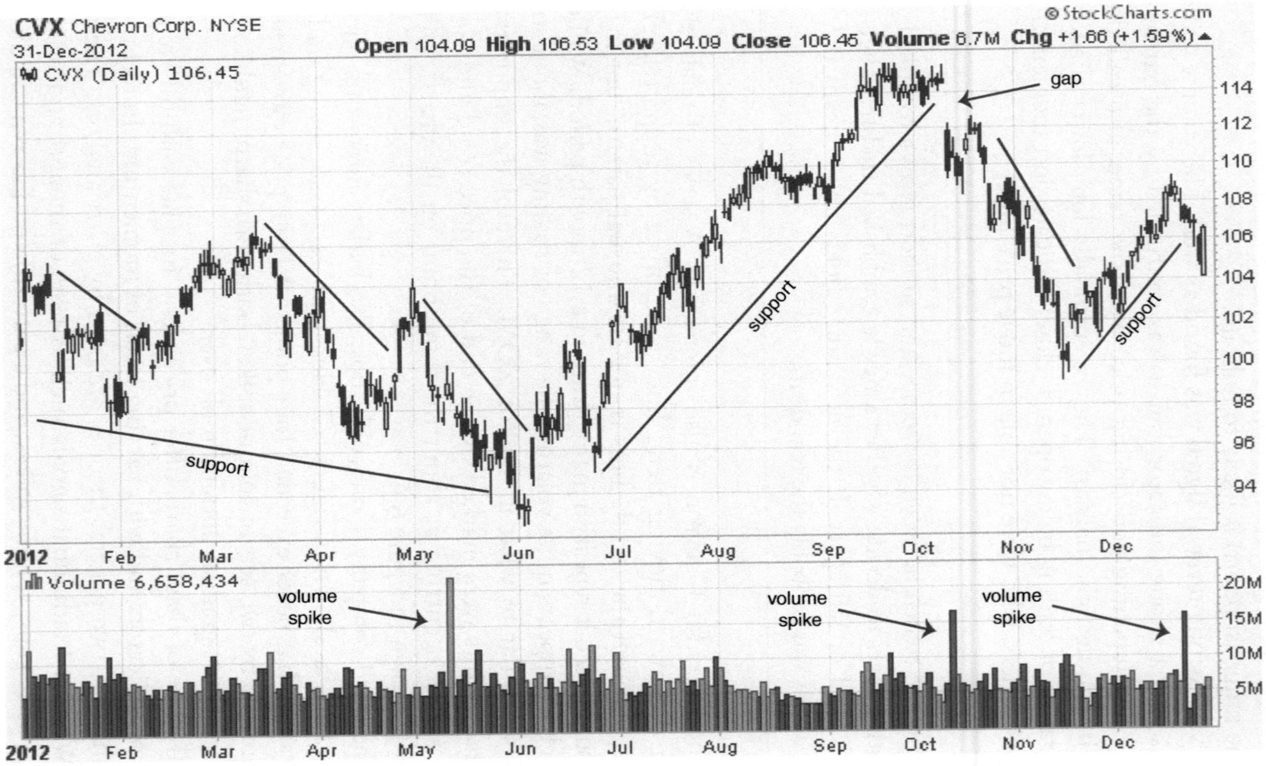

Figure 6.6 Chevron 1-year price chart

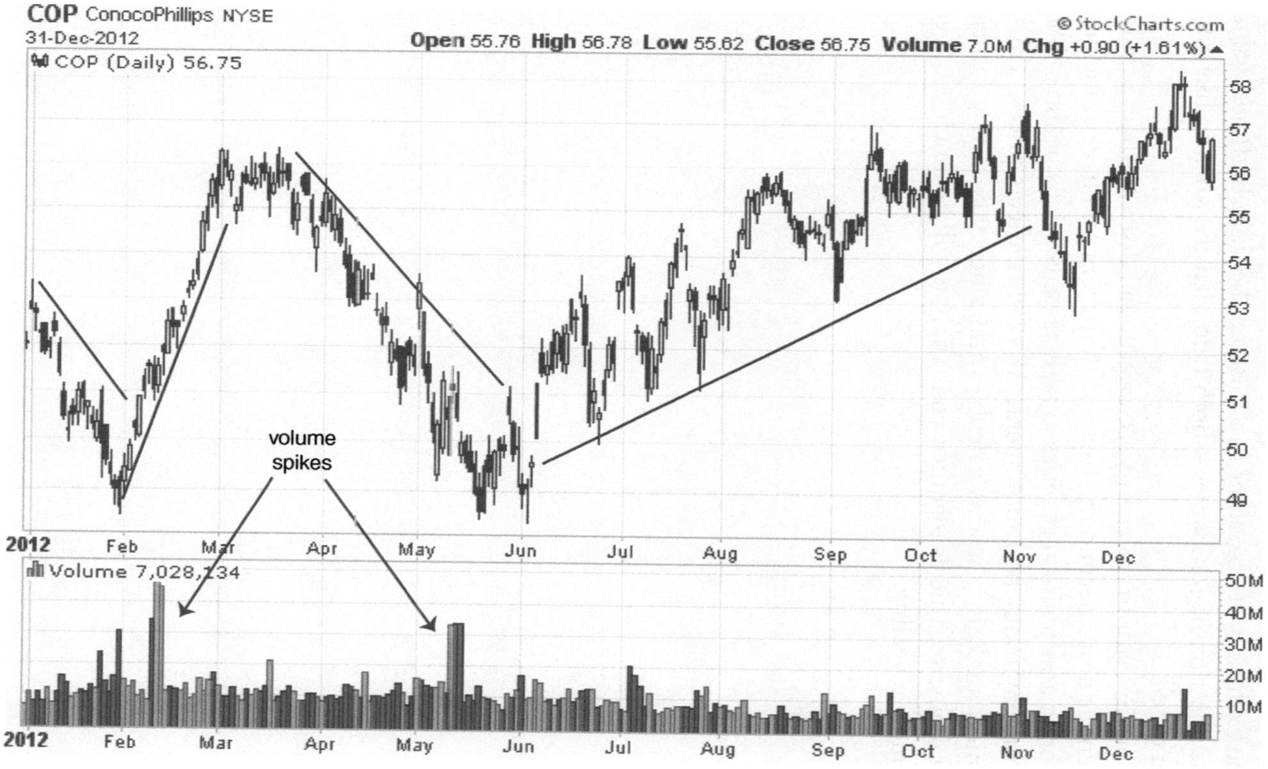

Figure 6.7 ConocoPhillips 1-year price chart

Fundamental analysis, the study of financial strength, profitability, and working capital trends and technical analysis (price and volume information) work well together. The tendency to rely on one to the exclusion of the other is a mistake, because both add value to the selection of companies—both help identify when to buy and when to sell.

As an investor in the energy sector, you will discover that the fundamentals move beyond the financial statement and have to include economic and political realities as well as perceptions of the public and the environmental movement. All of this affects prices. Thus, technical analysis is just as valuable, and when you review fundamentals and technicals together, your analysis will be much stronger and more revealing.

CHAPTER 7

ETFs and Index Funds

> For every human problem, there is a neat, plain solution—and it is always wrong.
>
> H. L. Mencken, *Prejudices: Second Series* (1920)

THE MUTUAL FUND INDUSTRY HAS BEEN AROUND NOT ONLY FOR decades but for centuries. The first known fund started in the Netherlands and was called *Eenddragt Maakt Magt* ("Unity Creates Strength"). This perfectly defines the concept of mutual funds, a pooling of funds for many individuals to create a larger, diversified portfolio.[1]

That first fund, begun by a merchant named Adriaan van Ketwich, was based on the idea that diversification was appealing to people with modest capital and that the fund would attract investors for that reason. A series of closed-end funds were set up in 1822 also in the Netherlands.[2]

In 1849, Switzerland began marketing investment trusts, and by 1880 Scotland had done the same. These were similar to the original concept of pooled and diversified funds as the model for what today is called the "mutual fund" or, more accurately, an "investment company." In the United States in 1893, the Boston Personal Property Trust was the first closed-end fund, and in 1907 the Philadelphia-based Alexander Fund was the initial instance of an open-end mutual fund (one allowing unlimited investment and dollar value). The modern mutual fund began in 1924 when Massachusetts Investors Trust began selling shares without limit as to the number of investors or the size of the portfolio. By 1928 the first no-load mutual fund was introduced, and the struggle between load and no-load continues to this day.[3]

The complexity of load fees and other fees has made mutual funds very complex. So many variations of fees are charged by modern mutual funds that a true comparison is very difficult.

Tip: To make valid comparisons between funds based on their fees, use the mutual fund cost calculator offered by the Securities and Exchange Commission (SEC). Also check the SEC page under "Mutual Fund Fees and Expenses" for a description of the many different fees mutual funds charge.

The modern mutual fund has been around in the United States for less than a hundred years, but today there are $26.8 trillion invested worldwide as of the end of 2012. Of this total, $14.7 trillion was invested in the United States ($13.0 trillion in mutual funds and $1.3 trillion in ETFs). Funds hold 28% of corporate equity in the United States, and 53.8 million households own shares of mutual funds.[4]

For decades in the modern mutual fund industry, many investors have discovered the convenience and risk reduction of the mutual fund alternative. However, if you want to focus on a specific industry, mutual fund investing presents problems as well. The typical fee for mutual funds is 2.5% per year versus only 0.44% on average for ETFs. While traditional mutual funds hold $9 trillion compared to only $1 trillion in ETFs, some clear differences have to be considered when comparing these two.[5]

One of the biggest problems with mutual funds is overdiversification. Fund managers will want to avoid owning too much stock in any one company, and so they are forced to diversify among many companies. As the asset value of a mutual fund grows, the need to spread out to an ever higher number of stocks makes the overall fund ineffective. Overdiversification turns a fund into a device with yields close to market averages; beating this average becomes more difficult as a fund grows.

Key point: An overdiversified fund—one so large it has to broaden its holdings—cannot beat the market because its portfolio is very much like the overall market.

Table 7.1 Four largest mutual fund returns, 2012

Symbol	Name	Assets	2012 Return
PTTAX	Pimco Total Return	$263 bil.	9.93%
VTSMX	Vanguard Total Stock Market Index Fund	190 bil.	16.25%
AGTHX	American Growth Fund of America	115 bil.	20.54%
VFINX	Vanguard 500 Index Investor Fund	111 bil.	15.82%
	S&P 500		**16.00%**

Sources: Bill Harris, "The 10 Biggest Mutual Funds: Are They Really Worth Your Money?" *Forbes*, August 8, 2012, at www.forbes.com and http:performance; Morningstar.com.

For example, four of the biggest mutual funds and their annual yields closely track the S&P 500, as shown in table 7.1.

The average returns of these four largest equity mutual funds was 15.635% for 2012, or about 0.365% *lower* than the market-wide return. In other words, size does not ensure higher returns; in fact, it often translates to lower returns. Even so, mutual funds have continued to grow over many decades and are still among the most popular ways to invest and save. Compared to direct ownership of stocks or bonds, mutual funds solve many problems and assign the decisions of what to buy or sell to professional management.

The landscape is changing rapidly, however.

Over the past two decades, a new type of mutual fund has emerged and is growing at a more rapid pace than the older style of mutual funds. The exchange-traded fund has grown from only two in 1995 to over 1,200 by the end of 2012. The growth curve of open-end mutual funds compared to ETFs is shown in figure 7.1.

The Differences between Mutual Funds and ETFs

A mutual fund collects investments from its members and then manages the overall funds in a portfolio. This portfolio is selected by professional management, and components are bought or sold based on management's opinion and expertise. For this a management fee is charged from each member as a percentage of the account's total value.

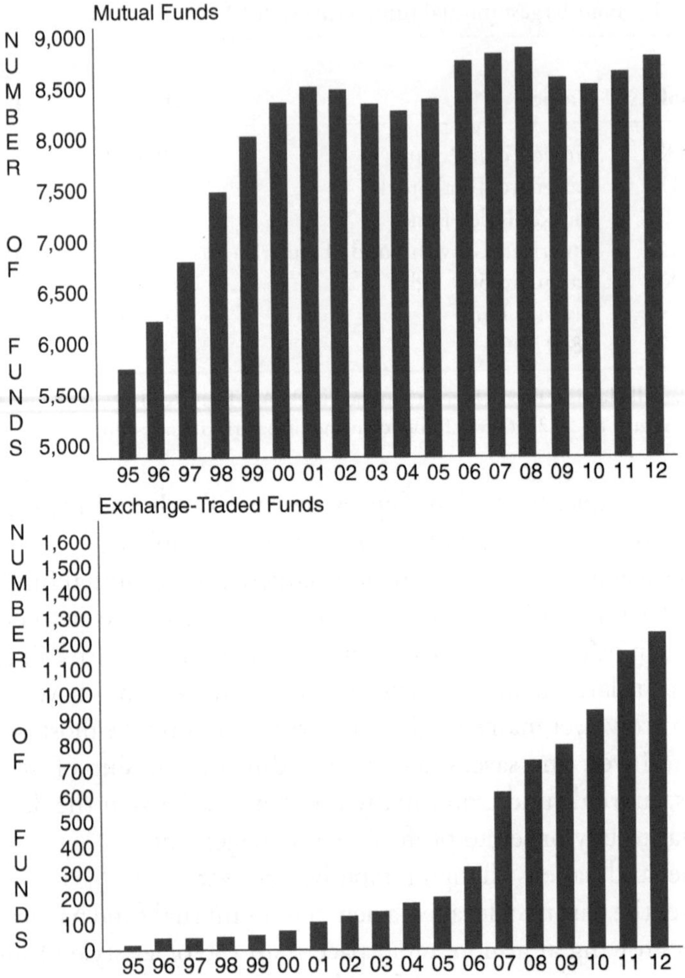

Figure 7.1 Number of mutual funds and ETFs, 18 years
Source: Compiled by author based on raw data atInvestment Company Institute (www .ici.org), "2013Investment Company Factbook"

An ETF does not require portfolio management because the portfolio consists of a predetermined "basket of securities." This basket only changes if one of the components is acquired, goes out of business, or is replaced for other reasons. That is, investors know in advance what their ETF buys and holds. Every ETF is identified by a specialty. This may be a particular sector, a country or region, or a type of product. For example, an equity ETF buys stocks, but a debt ETF specializes in bonds and other debts; a commodity ETF invests in a basket of commodities.[6]

Key point: The predetermined portfolio of an ETF allows investors not only to know what they get in advance but also to know that it won't change in the future.

Another distinction between the traditional open-end mutual fund and the ETF is the method of buying and selling shares and the time when those shares are assigned value. The mutual fund sets its net asset value (NAV) once per day, usually at 4 p.m. EST, the time markets are closed. The NAV consists of the market value for each of the fund's portfolio holdings. When individuals buy or sell shares, the price is always based on that end-of-day NAV. In an ETF, however, the price is continuously changing just like stock prices, and the ETF's value at any time during a trading day reflects the collective changes in its basket of securities.

There are many reasons that ETFs came into existence. One was the excessively high fees charged by many mutual funds. In addition to load fees, mutual funds charge a dizzying number of other fees. Among these, one of the more audacious is the 12b1 fee. Named for the SEC rule authorizing the fee, it was originally intended to help a mutual fund pay for its operating expenses, including advertising. When a fund charges a 12b-1 fee, it could use the money to advertise for new subscribers. Investors may be confused, however, because even when picking a no-load fund, they might end up being charged a 12b-1 fee and many other fees as well. The level of the 12b-1 fee usually runs at 0.25% of each account's value, assessed each year.[7]

Unlike the mutual fund with its array of fees, an ETF is more likely to limit its management fees since no portfolio decisions are required. For example, a mutual fund investor may pay as much as 1.67% or more in fees, but an ETF investor is likely to be charged only 0.4% on average.[8]

What Are Energy Sector Exchange-Traded Funds?

ETFs in the energy sector include dozens of highly focused funds, with holdings in oil and gas and alternative energy. Making choices even richer, many are leveraged, described as 2x and 3x (meaning movement in the ETF exceeds movement in the securities by two times or three times, higher leverage, and higher risk/opportunity). Traders can also pick bearish ETFs, which increase in value when the securities in the ETF fall.

Key point: A leveraged fund grows profits more quickly than unleveraged ones, but it can also grow losses more quickly.

ETFs are very flexible and the choices are varied, even more so than in traditional mutual funds. ETFs also allow options trading in many instances, thus increasing leverage possibilities even more. Instead of trading shares of an ETF, traders can buy or sell options to maximize either bullish or bearish movement in the entire basket of securities. Traditional mutual funds do not allow options trading, which explains why ETFs have become so popular so quickly. While ETFs are still outnumbered by traditional funds, the picture is changing quickly.

The most popular energy ETFs focus on oil and gas. These are the most active commodities and also the most popular class of ETFs and ETNs. A list of these includes the following:

BARL—Morgan Stanley S&P 500 Crude Oil Linked ETN
BNO—United States Brent Oil Fund
CRUD—Teucrium WTI Crude Oil Fund
DBO—PowerShares DB Oil ETF
HUC—Horizons BetaPro Winter-Term NYMEX Crude Oil ETF
FRAK—Market Vectors Unconventional Oil and Gas ETF
IOIL—Global Crude Oil Small Cap Equity ETF
OIH—Market Vectors Oil Services ETF
OIL—Goldman Sachs Crude Oil Total Return ETN
OILZ—ETRACS Oil Futures Contango ETN
OLEM—iPath Pure Beta Crude Oil ETN
OLO—PowerShares DB Crude Oil Long ETN
SCO—ProShares UltraShort DJ-AIG Crude Oil ETF
SNDS—Sustainable North American Oil Sands ETF
UCO—ProShares Ultra DJ-AIG Crude Oil ETF
UHN—United States Heating Oil Fund ETF
USL—United States 12 Month Oil Fund ETF

USO—United States Oil Fund ETF
WCAT—Jefferies TR/J CRB Wildcatters Exploration and Production ETF

Short[9]
DTO—PowerShares DB Crude Oil Double Short ETN
DDG—The Short Oil and Gas ProShares ETF
DNO—United States Short Oil Fund ETF
DUG—UltraShort Oil & Gas ProShares ETF
SZO—PowerShares DB Crude Oil Short ETN

Leveraged[10]
FOL—FactorShares Oil Bull / S&P 500 Bear ETF—**2X**

ETFs can be defined in several ways. For example, they may be set up as single contracts or multiple contracts. A single-contract ETF includes a basket of securities in only one form of energy; a multi-contract ETF is broader and may include several different energy types or a combination of energy and non-energy commodities.

Many ETFs are bearish, which means the value of shares grows if and when the value of the basket of securities declines. . For many investors, shorting an ETF is just as risky and complex as shorting stock; a bearish ETF offers an alternative to this approach. Buying shares of a bearish ETF is equivalent to shorting a position; but it is less risky because the share value is long.

In addition to oil and gas ETFs, there are many other specialized ETFs to suit the investment goals of anyone interested in other forms of energy. For example, several ETFs focus on clean and alternative energy:

FAN—First Trust ISE Global Wind Energy Index Fund
GEX—Market Vectors Global Alternative Energy ETF
ICLN—iShares S&P Global Clean Energy Index Fund
KWT—Market Vectors Solar Energy ETF
PBD—PowerShares Global Clean Energy Portfolio ETF
PBW—PowerShares Wilder Hill Clean Energy ETF
PWND—PowerShares Global Wind Energy Portfolio
PZD—Power Shares CleanTech Portfolio ETF
QCLN—First Trust NASDAQ Clean Edge Green Energy ETF
TAN—Guggenheim Solar ETF

Nuclear energy is yet another specialized form of energy included in energy -related ETFs. In this subgroup are the following ETFs:

NLR—Market Vectors Nuclear Energy ETF
NUCL—iShares S&P Global Nuclear Energy Index ETF

Another specialized version of the ETF focuses on natural gas:

DCNG—iPath Seasonal Natural Gas ETN
FCG—First Trust ISE-Revere Natural Gas ETF
BOIL—ProShares Ultra DJ-UBS Natural Gas ETF
GASZ—ETRACS Natural Gas Futures Contango ETN
GAZ—iPath DJ AIG Natural Gas TR Sub-Index ETN
MLPG—UBS E-TRACS Alerian Natural Gas MLP Index ETN
NAGS—Teucrium Natural Gas Fund
UNG—United States Natural Gas ETF

Short:
DDG—The Short Oil and Gas ProShares ETF
KOLD—UltraShort DJ-UBS Natural Gas ETF

Leveraged:
GASL—Direxion Daily Natural Gas Related Bull Shares ETF—**3X**
GASX—Direxion Daily Natural Gas Related Bear Shares ETF—3X[11]

In addition to the specialized varieties of ETFs, some track the entire energy sector, representing a basket of securities across the entire sector, such as the following:

AXEN—ishares MSCI ACWI ex US Energy Sector Index ETF
AXEN—MSCI ACWI ex US Energy Sector Index Fund
CHIE—China Energy ETF
DBE—PowerShares DB Energy ETF
EMEY—MSCI Emerging Markets Energy Sector Capped Index Fund
EMLP—First Trust North American Energy Infrastructure ETF
ENY—Canadian Energy Income ETF
FILL—MSCI Global Energy Producers Fund
FXN—Energy AlphaDEX Fund

ICLN—iShares S&P Global Clean Energy Index ETF
IPW—SPDR S&P International Energy Sector ETF
IXC—S&P Global Energy Index Fund
IYE—Dow Jones U.S. Energy Sector Fund
JJE—iPath DJ-UBS Energy Total Return Sub-Index ETN
OGEM—Energy GEMS ETF
ONG—iPath Pure Beta Energy ETN
PSCE—S&P SmallCap Energy Portfolio
PUW—PowerShares Wilder Hill Progressive Energy ETF
PXE—PowerShares Dynamic Energy Exploration and Production ETF
PXI—Dynamic Energy
RGRE—RBS Rogers Enhanced Energy ETN
RJN—ELEMENTS Rogers International Commodity Energy ETN
RYE—S&P Equal Weight Energy ETF
UBN—UBS E-TRACS CMCI Energy Total Return ETN
VDE—Energy ETF
XLE—Energy Select Sector SPDR

Leveraged:
ERX—Daily Energy Bull 3X Shares
ERY—Daily Energy Bear 3X Shares

The number of ETFs and specializations of them grows every year, and the field can be subdivided in several ways: by specific energy subsector, by market-wide coverage, diversification between energy and other commodities, for example. ETFs further are distinguished by being bullish or bearish and by their leveraged posture in the market.

In addition to ETFs, commodity index funds provide another way of investing in the energy market as well as in other commodities markets.

Commodity Index Funds

Commodity-based index funds include energy as well as other forms of commodities. Compared to other methods for taking positions in the futures market, index funds are probably the most economical and most diversified choice. These funds heavily favor energy commodities, and this reflects the importance of energy commodities as the dominant segment of the market.

Key point: A big draw of commodity index funds is the built-in diversification among energy and other types of commodities. This cannot be accomplished as economically in any other way.

Each of these funds invests in instruments for specific commodities or in derivatives that track movement in commodity values. The first such fund was the Commodity Research Bureau (CRB), founded in 1958 and subsequently merged with Thomson Reuters and Jefferies (the Reuters/Jefferies CRB Index). In 1991, Goldman Sachs formed its own Goldman Sachs Commodity Index (GSCI). The Dow Jones AIG Commodity Index controls the weight of each components allowed; following AIG's financial liquidity problems in 2008, this fund was sold to UBS (Union Bank of Switzerland) and today is called the Dow Jones-UBS Commodity Index. Other commodity index funds are operated by Standard & Poor's, known as the S&P World Commodity Index, and by the Rogers Index, known as the Rogers International Commodity Index (RICI).

These commodity index funds tend to emphasize energy over other classes of commodities (metals, agriculture, and livestock). The Reuters/Jefferies is the only one of the five funds that places more weight on agricultural than on energy futures. The breakdown is summarized in figure 7.2.

The breakdown of commodities held by the Dow Jones-UBM Commodity Index is summarized in figure 7.3.

The S&P World Commodity Index division of holdings is shown in figure 7.4.

And the RICI holdings are summarized in figure 7.5.

Holdings of the GSCI are summarized in figure 7.6.

The primary advantage of using ETFs and funds over direct trading in either futures or stocks is that this lowers the cost to trade and also lowers the minimum investment requirement. ETFs and index funds trade on public stock exchanges and are very liquid. In comparison, the cost of buying and selling futures contracts is much higher, as are the risks.

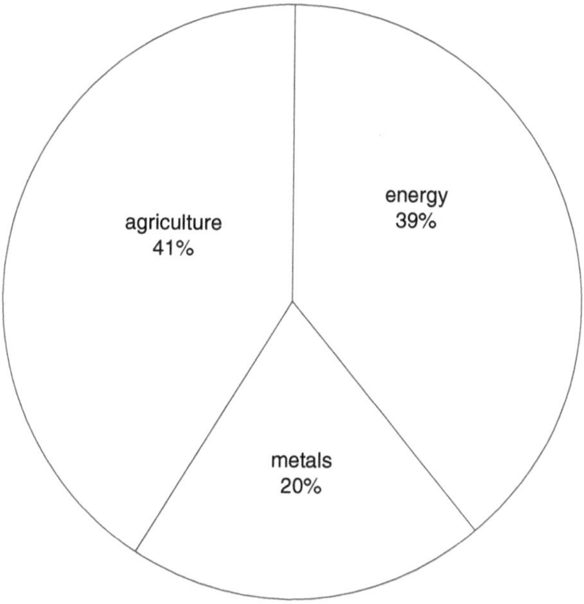

Figure 7.2 Reuters/Jefferies CRB Index
Source: prepared by author from raw data at "Thomson/Reuters / Jefferies CRB Index-Calculation Supplement," May, 2012, at http://thomsonreuters.com

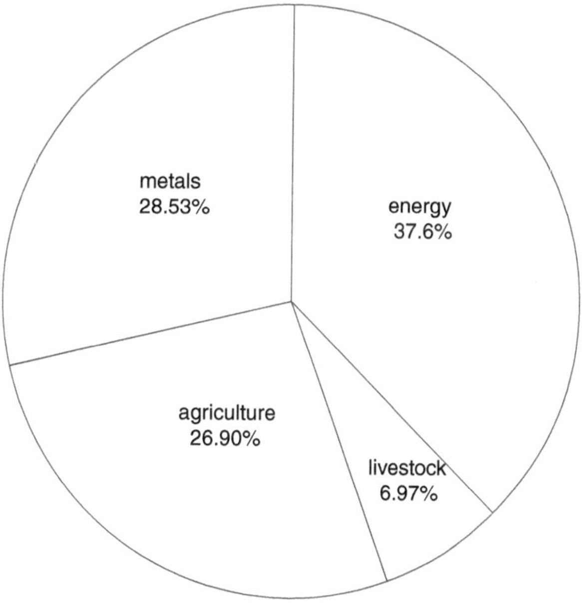

Figure 7.3 Dow Jones-UBS Commodity Index
Source: prepared by author from raw data at www.djindexes.com/commodity, downloaded May 3, 2013

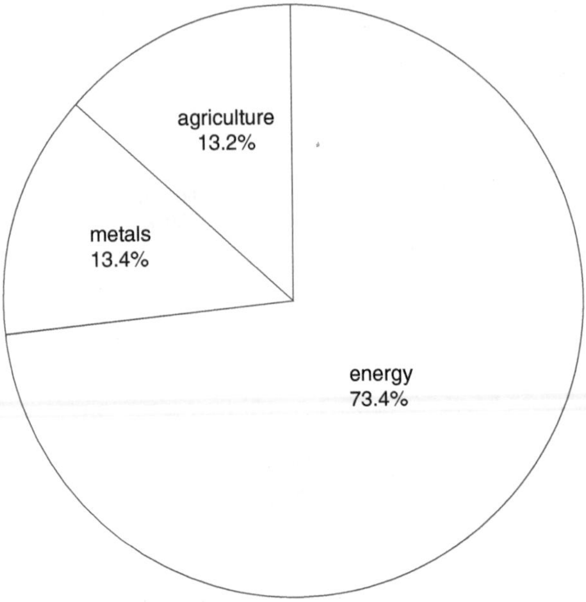

Figure 7.4 S&P World Commodity Index
Source: prepared by author from raw data at "S&P World Commodity Index (www.standardandpoors.com), downloaded May 3, 2013

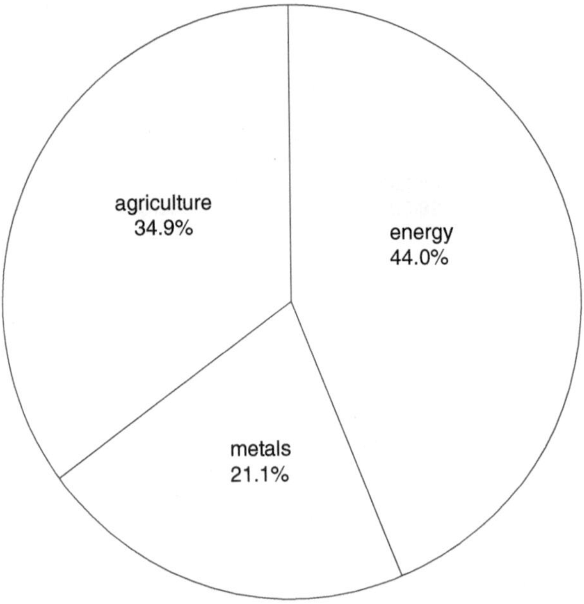

Figure 7.5 Rogers International Commodity Index (RICI)
Source: prepared by author from raw data at www.rogersrawmaterials.com -downloaded May 3, 2013

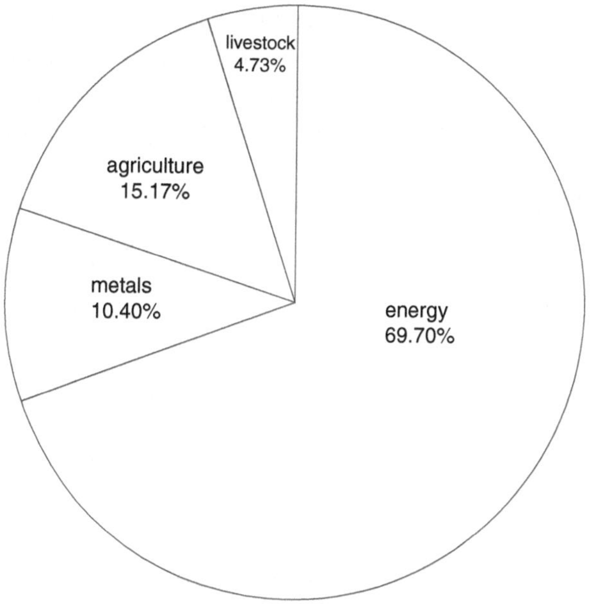

Figure 7.6 Goldman Sachs Commodity Index (GSCI)
Source: prepared by author from raw data at "S&P Dow Jones Indices: S&P GSCI Methodology," January 2013 (www.spindices.com)

Risks and Rewards

Mutual funds have traditionally been known as convenient, diversified, and professionally managed. ETFs are in a sense less diversified because the basket of securities shares common traits, including vulnerability to the same market forces. This is a risk. However, it is also convenient to be able to invest in an entire sector instead of needing to pick one stock out of many.

Key point: ETFs are less diversified than other mutual funds, and fundamental analysis of them is difficult because the ETF is a collection of securities, each with its own, often conflicting, set of fundamentals.

Risk usually translates to degrees of volatility. Higher volatility means higher risks. Diversification is supposed to offset risk. If you put all of your investment capital into a single product and it loses value, your entire portfolio suffers. This is why most financial professionals tell their clients to diversify, not only among different products but also among different types of products (asset allocation).

The ETF diversifies among several products that share common traits (same industry, product, or country, for example). An energy ETF continues to face volatility; if the entire industry (or subsector of it) suffers a setback in competitiveness, it's likely that all the stocks in that basket of securities are going to move in the same direction. Some ETFs are not really diversified at all. For example, one of the better-known oil and gas ETFs is United States Oil Fund (USO), and it is invested 100% in sweet light crude oil. Its prospectus explains:

> The investment objective of USOF is for the daily changes in percentage terms of its units' NAV to reflect the daily changes in percentage terms of the spot price of light, sweet crude oil delivered to Cushing, Oklahoma, as measured by the daily changes in the price of the futures contract on light, sweet crude oil as traded on the New York Mercantile Exchange (the NYMEX) that is the near month contract to expire, except when the near month contract is within two weeks of expiration, in which case it will be measured by the futures contract that is the next month contract to expire (the "Benchmark Oil Futures Contract"), less USOF's expenses.[12]

The potential for holdings of multiple delivery dates and even of separate crude oil futures is allowed for in this prospectus description; even so, focus of the fund is on one type of energy future, and thus it is diversified only to the extent that multiple contract positions are possible. However, for ETF investors, the convenience of investing in crude oil energy futures through an ETF may outweigh questions about diversification. USO, like other ETFs, is traded on the exchanges just like stock and for transaction costs equivalent to buying and selling shares of stock. It is listed on the NASDAQ, and an additional attraction of this method for trading futures is that you can buy and sell options on USO as well.

For example, as of the beginning of May 2013, USO was trading at \$34.16. You could buy or sell May 2013 calls with 34.50 strike at 0.41 (\$41) or puts at 0.78 (\$78). The June options for the 34.50 strike were 0.91 (calls) and 1.28 (\$128) for puts.[13]

That is, a trader could buy a call in the belief that USO's market price would rise or buy a put in the belief that the price would fall. Dozens of combination strategies can also be employed, so that options represent a very low-cost method for trading in the ETF market. Every option controls 100 shares of stock, and movement in the option value enables a trader to duplicate that price change for a small fraction of the cost of 100 shares. For example, the May 34.50 calls were valued at 1.2% of the price of 100 shares. However, because these were set to expire two weeks later, the value has to be judged based on that as well as on the proximity factor (between the option's strike and current price per share). The point is that options based on an oil and gas ETF present yet another method for investing in the energy market.

Key point: The leverage possible with options allows you to trade in energy ETFs, an effective combination of built-in diversification and leverage.

The question of risk associated with diversification is one that may attract you to one ETF over another. If an ETF is restricted to one segment of the energy market only, a volatile condition makes it more risky than an ETF that has greater diversification or a commodity index fund with several different types of commodities. For example, the GSCI commodity fund includes about 69% energy investments. This consists of crude oil (38%), Brent crude oil (13%), RBOB Gas (4%), heating oil (4%), GasOil (5%), and Natural Gas (5%)—plus another 31% in metals, agriculture, and livestock. This is effective diversification among commodities.

Within the ETF market, even oil and gas funds may be more diversified than USO and some others. For example, WCAT (another oil and gas ETF) is very diversified, with just under 37% of its portfolio in ten positions and the remaining 63% in other holdings below 2.5% of the total. Table 7.2 summarizes the top ten holdings of this ETF.

Table 7.2 Top fund holdings for WCAT

Name	% of Total
PetroBakken Energy Ltd	5.642%
Rosetta Resources Inc	5.371%
Energy XXI Bermuda Ltd	5.263%
Gran Tierra Energy Inc	3.392%
Vanguard Natural Resources LLC	3.044%
Forest Oil Corp	3.038%
Petrobank Energy & Resources L	2.982%
Petrominerales Ltd	2.837%
Celtic Exploration Ltd	2.619%
Stone Energy Corp	2.607%

Source: www.bloomberg.com, downloaded May 3, 2013.

How to Use ETFs

Most investors buy shares of ETFs as part of a balanced portfolio. They will own some stocks as well as shares of an ETF. Thus, if your portfolio includes energy ETFs, you have a portion of your capital in diversified energy companies or in futures, but the ETF has the added advantage that it trades like stocks.

A number of other investors tend to focus on ETFs or index funds as a way to diversify within a broader market and as an alternative to buying energy stocks or commodities directly. In this method, the ETF portion may be combined with commodity index funds or direct ownership in energy stocks or even in energy futures.

Key point: Some interesting variations on the idea of diversification can be put into practice when combining energy ETFs and commodity index funds.

Given the broad range of energy-specific and other ETFs, it has now become both possible and practical to construct a diversified portfolio entirely out of ETFs, index funds, and mutual funds. Dividing and diversifying by capitalization size, geography, products, futures, and stocks is much easier today than in the past; for energy investing, selecting ETFs to represent a range of different energy stocks or futures is easy—even including some positions leveraged with the use of options (see chapter 8).

ETFs are used in energy investing for several advantages, such as the following:

Lower costs: This is possible with ETFs and index funds, especially compared to the costs of trading futures directly. If you try to duplicate the ETF portfolio by taking up positions in some of its components, you will end up paying much more in transaction costs than you would be just trading shares of the ETF.

Lower risks: The built-in management of a series of positions invariably has the effect of reducing overall volatility. Positions with higher volatility will be evened out by those with lower volatility, which means that the entire portfolio tends to provide the same types of products but with far lower volatility.

Diversification: This is a double-edged sword. An ETF does diversify among the defined types of products (crude oil, for example), but if the crude oil market gets bad news, it's likely that all of the stocks or futures in the basket of securities will lose value in the same way and at the same time. The principle of diversification is to avoid this. One solution is to buy shares not only in one ETF but in several, diversifying by market sector, product type, and even by country or region.

Long or short, bull or bear: Most traders do not want to short stock, but the ETF solves this problem. When you buy a short or bearish ETF, portfolio values rise when the value of the basket of securities falls. This allows you to take up a short position in the securities, while avoiding the high risks and costs of shorting the securities.

Leverage: Some ETFs are labeled as 2X, which means that the value changes at twice the rate of change in the basket of securities; others are 3X, which means three times more movement. However, the opportunity is high and so is the risk. You can make money more quickly, but you can also lose money more quickly in these leveraged ETFs.

Another way you can enrich your portfolio by using ETFs is to combine the three different product types. The ETF is associated with a basket of securities consisting of stocks; the ETB is a debt-based version, holding positions in bonds or money market securities.

And the ETC is for commodities. All of these are referred to broadly as ETFs, and the use of the other designations is inconsistent. But knowing that you can use ETFs to allocate as well as to diversify—and at very low trading fees—is quite an advantage. Although financial professionals working on a fee basis might not like it to be widely known, the ETF and index fund world largely eliminate the sales loads of traditional mutual funds and make trading as easy as trading in shares of stock.

What about Profitability?

Every investor wants to pick and time purchases and sales of products to maximize profits. In the ETF market, timing is based on how the market is performing and on whether a particular sector is prominent and on the upswing. If the sector is profitable, holding ETF shares is also likely to track that uptrend, and shares will appreciate in market value. But you need to rely on your own judgment in deciding when a particular sector is most likely to begin moving in the direction you desire. This is not to say you have to always trade the energy market in short-term buy or sell decisions, but being aware of how prices move will help you time your investment decisions for best results.

Some market advisers will tell you that you cannot make money consistently, but the makeup of the ETF accomplishes a specific attribute, with the securities sharing common attributes such as sector, risks, and opportunities. By combining a basket of securities with common elements, you (a) limit the potential losses resulting from declines in any one of the securities, and you also (b) limit the profit potential you would gain when one or two of those securities rises in value. Considering the inherent market risk of the energy sector and its volatility, this evening out of the opportunity and risk exposure is desirable in one sense, because the ETF shares will tend to move together. But in another sense, it is not true diversification because all of the components in that basket—since they share the same attributes—are vulnerable to the same market risks.

The underlying principle of the ETF is that two related products (like stocks in the energy sector) tend to move in the same direction at the same time. This correlation means that when the market is bullish, the tendency is for all (or most) of the basket of securities to gain in value. However, you are not exposed to the risk of an undiversified portfolio, which can lose value just as quickly and due to the same tendency (correlation).

Key point: It often is true that related securities move in the same direction at the same time. This is a great advantage to ETF investing. It may also be a great disadvantage.

Another aspect of ETF trading looks beyond this risk-versus-opportunity consideration. Most stock investors understand that dividends represent a significant portion of total return (and often most of that return). One criticism of the ETF market is that in buying the basket of securities, you give up the dividend. However, some ETFs actually focus on dividends and pay them to shareholders. For example, the Energy Select Sector SPDR (XLE) yields about 1.81% as of May 2013, and its holdings are 96.96% in oil and gas investments (the remainder is divided between utilities and basic materials). The top ten holdings of XLE are shown in table 7.3.

The dividend of 1.81% is not impressive compared to what some individual energy companies pay. Another dividend-yielding ETF in the energy sector is the Guggenheim Canadian Energy Income Index (ENY). As of the beginning of May 2013, this ETF was yielding 0.77%. The top ten holdings are summarized in table 7.4.

Table 7.3 XLE—Energy Select Sector SPDR—Top 10 Holdings

	Company Name	% of Holdings
1	Exxon Mobil Corp ORD	17.15%
2	Chevron Corp ORD	14.84%
3	Schlumberger NV ORD	6.53%
4	Occidental Petroleum Corp ORD	3.47%
5	ConocoPhillips ORD	3.41%
6	Anadarko Petroleum Corp ORD	3.21%
7	Halliburton Co ORD	2.84%
8	Phillips 66 ORD	2.74%
9	EOG Resources Inc ORD	2.67%
10	National Oilwell Varco Inc ORD	2.41%

Source: www.thestreet.com, May 3, 2013.

The dividend of 0.77% was still low compared to what some other companies in the sector pay. You can average a much greater level of dividend income by directly buying a few stocks. For many traders, capital limitations also limit diversification, and holding only a few company stocks is not adequate diversification. For others, a higher dividend yield is a worthwhile trade-off for the risks involved. Table 7.5 presents four examples of high-yielding companies in the energy sector.

These four companies average 4.65%, an attractive yield in the current market of low interest yield (as of the beginning of May 2013). Gaining nearly 5% per year is appealing, and if you were to buy shares of stock in these companies and reinvest quarterly dividends, your annual yield would be 4.7%.[14] You could create your own diversified portfolio of energy stocks that would work just like an ETF with equal numbers of shares in each of these companies if your purpose was to focus on high yield while also participating in

Table 7.4 ENY—Guggenheim Canadian Energy Income Index—Top 10 Holdings

	Company Name	% of Holdings
1	Suncor Energy Inc ORD	8.04%
2	Canadian Oil Sands Ltd ORD	7.37%
3	Cenovus Energy Inc ORD	6.44%
4	Imperial Oil Ltd ORD	6.29%
5	Meg Energy Corp ORD	5.83%
6	Athabasca Oil Corp ORD	5.77%
7	Canadian Natural Resources Ltd ORD	5.76%
8	Baytex Energy Corp ORD	4.67%
9	Petrobank Energy And Resources Ltd ORD	4.52%
10	Southern Pacific Resource Corp ORD	4.49%

Source: www.thestreet.com, May 3, 2013.

Table 7.5 Energy company dividend yields

PBR—Petro Brasileiro S.A.	6.3%
BP—British Petroleum	4.8
COP—ConocoPhillips	4.3
CVX—Chevron	3.2

Source: Charles Schwab & Co., downloaded May 3, 2013.

the energy sector. Considering the relatively low cost per transaction when using a discount broker, this is not an expensive alternative to the ETF if you want to maximize dividend yield among energy stocks.

> **Tip:** If the ETF does not appeal to you, consider building your own basket of securities based on similar market attributes and high dividend yield.

In other words, the ETF is convenient, but it will not always meet an individual's investing needs. It could be worth the initial higher transaction cost to create your own ETF-like portfolio and gain 5% per year. (And, as the next chapter demonstrates, you could vastly increase those returns with some conservative options strategies. In that case, you create the potential of gains from three sources: dividends, option premium, and capital gains on the stock.) This idea is revisited in chapter 8 and explained in detail based on the above-mentioned companies.

If you prefer to trade in the debt market, direct ownership of bonds or money market securities is expensive and may also require more capital than many traders can afford to place at risk. The ETN (exchange-traded note) is a basket of debt securities within one sector, country, or region. For example, an energy sector ETN will hold debt securities for a range of energy companies, which is ideal because it provides diversification combined with low cost.

The next chapter demonstrates how you can expand an equity position in the energy market with options. This can be accomplished in one of two ways: by trading options on energy stocks or by trading options on futures (or futures ETFs). The amazing leverage of options is both appealing and risky, but many conservative options strategies make this a market worth study and consideration. You may be able to expand your futures, stock, index fund, or ETF trades into leveraged and profitable portfolio positions, using options to create income and to hedge market risks.

CHAPTER 8

OPTIONS

> No one can be a great thinker who does not recognize that as a thinker, it is his first duty to follow his intellect to whatever conclusions it may lead.
>
> John Stuart Mill, *On Liberty*, 1859

SINCE THE START OF THE TWENTY-FIRST CENTURY, DERIVATIVE activity has grown substantially. Why?[1]

The futures market is one form of derivative contract. A futures contract leverages the value of an underlying commodity. When you trade an option on a futures contract, you create extreme leverage, a form of "leverage on leverage." The option's value changes as the futures contract changes, and the futures contract moves with the value of the underlying commodity.

For an individual trader, options can be bought or sold on stock or on futures. For the purpose of explaining options and how they work, this chapter describes the most popular version of the market, which is the trading of options on stock. In the energy market, then, shares of energy stocks (or ETFs) can be bought or sold, or options on those underlying assets can be traded for a small fraction of the cost.

For institutions, where the volume of trading is significantly higher than that of individuals, a growth in derivatives has also occurred in the first decade of the current century. The causes of this expanded market are complex, but both institutions and individuals have come to recognize the value of the options market. In the past, it was viewed as a highly speculative and risky type of trading. Today, options are used more to hedge risks in an equity portfolio, and many highly conservative strategies may further enhance income while protecting against market risks. The institutional side of options trading in recent years was summarized in a paper published in 2011:

> During the past decade, many institutional portfolio managers added commodity derivatives as an asset class to their portfolios. This addition was part of a larger shift in portfolio strategy away from traditional equity investment and toward derivatives based on assets such as real estate and commodities. Institutional investors' use of commodity futures to hedge against stock market risk is a relatively recent phenomenon. Trading in commodity derivatives also increased along with the rapid expansion of trading in all derivative markets.[2]

The same assumptions are applied to individuals. When you use options to hedge stock market risk in a variety of ways, you make efficient use of options (as one form of derivative) rather than the alternative use, speculating in search of fast profits.

This chapter explains how the options market works and, specifically, how a portfolio of energy stocks may be protected with the use of options, not only to guard against market risks but also to expand current income. In such a portfolio, it is possible to generate profits from three sources: capital gains on fundamentally well-selected stocks, dividend yield, and option premium.

What Exactly Are Options?

Options are intangible contracts that allow traders to buy or sell an underlying security (shares of stock or contracts on futures, for example). Every option on stock refers to and controls 100 shares and is identified with "standardized terms"—the type of option (call or put), the underlying security, a strike price, and an expiration date. These terms cannot be changed.

> **Key point:** Every option controls the price movement of 100 shares of an underlying stock and can be bought for a small fraction of the cost of 100 shares.

Calls give their owners the right (but not the obligation) to buy 100 shares at a fixed price. Puts give their owners the right (but not the obligation) to sell 100 shares. The underlying security is the stock, ETF, index, or futures contract. Every option expires in the future, and after expiration options become worthless. The strike price is the fixed price at which the call owner can buy or the put seller can sell, no matter how high or low the market price has moved.

For example, you buy a 43 call at a point when the stock is valued at $44.03 and pay 1.64 (in options listings, the strike is expressed without dollar signs, so 43 translates to a strike of $43 per share; the price of the option, called its premium, is expressed on a per-share basis, but since each option controls 100 shares, 1.64 translates to $164). The option expires in 75 days. You hope that the underlying price will rise far enough and fast enough so that the 43 call will end up worth more than the $164 paid. At expiration in 75 days, if the stock's price is $44.64, that is the breakeven price.

Tip: Think of options as a means for fixing the price of 100 shares. You do not have to buy or sell that stock, but you can if and when price movement is favorable.

Another example: You buy an 18 put for 0.51 ($51) on a stock currently worth $18.86. The put expires in 47 days. You hope that the price per share *declines* below the strike of 18 ($18 per share) within the 47 days before expiration. So breakeven is $17.49 per share (18 strike minus 0.51 paid for the option). If the price declines below that level, you can sell the put at a profit; as long as it remains at $17.49 or above, the put will expire worthless.

In these basic examples of purchases of either a call or a put, you enter into a position in the belief that the stock price will move in the desired direction (upward for calls or downward for puts). The challenge is that the stock has to move far enough within a limited time in order for the option to become profitable. In addition to time working against the long-side trader, *time value* also poses a problem. As expiration approaches, time value declines, and by the expiration date it is at zero.

Key point: Time works against an option buyer, because value declines as expiration approaches. For an option *seller*, this is a big advantage.

In order for a long option to become valuable, it has to move *in the money* (meaning the price per share moves higher than a call's strike or lower than a put's strike). Intrinsic value is equal to the number of points in the money. For example, if the strike of a call is 43 and the price per share is $43.75, it is three-quarters of one point (0.75) in the money. If a put's strike is 18 and the underlying stock is worth $19 per share, the put is 1.00 ($100) in the money. But if the put moves above $18 per share, it moves out of the money, and intrinsic value is at zero.

Who Are the Players?

Every options transaction involves a buyer and a seller. The Options Clearing Corporation (OCC) facilitates the market by matching up sellers and buyers. Trades are placed through a broker and executed on an options exchange or one of the stock exchanges.

When you buy an option (go "long"), you begin with a "buy to open" trade; when you sell, you enter a "sell to close" trade. However, a long option is not always sold. It can also be exercised. When you exercise the option, you claim your right to buy 100 shares at a call's strike (you "call" the shares) or to sell 1200 shares at a put's strike (you put the 100 shares to the seller). Either option may also expire worthless.

When you sell an option (go "short"), you begin with a "sell to open" trade; when you close, it is a "buy to close" trade. Shorting options involves a sequence opposite of the well-known "buy, hold, sell" of the long side. Sellers experience the sequence of "sell, hold, buy." When the trade is opened, you as a seller receive the premium value of the option, and that is yours to keep no matter what happens next. However, in exchange for that income, you grant rights to a buyer. You could have 100 shares called away at the strike, or you could have 100 shares put to you at the strike.

For example, you sell a 17 call for 2.33 and you receive $233. At that time the stock is worth $18.86 per share, and this call has 75 days to go before expiration. As a seller, you find time is on your side. Time value declines as expiration approaches, and this means the option is worth less. You hope that the short call will be worth less than the $233 you received. In that case, you can close and take a profit. If the call's value falls to $150, you can sell and take a profit of $83 ($233 received less $150 paid to close). If the stock remains above the 17 strike, the short call will be exercised. For example, on the last trading day, the stock price is $23 per share. Your 17 call is exercised, and you are required to deliver 100 shares at $17 per share. Because the current price is $23, this is a loss of $600. However, you originally received $233, so your net loss is reduced to $367 ($600—$233). In this case, your breakeven price was $19.33 per share (strike of 17 plus 2.33 for the option).

Another example: You sell a 43 put for 0.84 ($84) when the price of the stock is $44.03. There are 47 days to go before expiration. As a seller, you hope the stock's price remains above $43 per share (out of the money). If the stock price falls below the 43 strike by the expiration date, you are required to deliver 100 shares at $43 per share. But because you received $84 when you sold the call, your net loss is reduced by the $84 you received for selling the option.

Key point: Using only single options to control 1,200 shares of stock is only a starting point. Many combinations of calls and puts, both long and short, open up the possibilities for dozens of strategies. These may lead to profits in markets moving up, down, or sideways.

Reviewing single option positions, either long or short, is only a starting point. Dozens of advanced strategies involve buying or selling combinations of calls or puts at the same strike (called a straddle) or with one above and the other below the current price (a spread).

Option Terminology

The options market has its own jargon, and anyone interested in entering the market for the first time will have to master the terms as well as develop an understanding of the risks involved with each type of options trade.

The *call* provides its buyer with the right to buy 100 shares at a fixed price before the expiration date. The *put* gives its buyer the right to sell 100 shares at a fixed price.

The *strike* is the fixed price per share for every option, expressed in value per share. For example, the value or *premium* of an option is reported at 1.64, which means its current premium is $164. An option buyer may exercise, sell at a profit, or allow the contract to expire worthless.

The *expiration* is the date when every option becomes worthless. It is expressed by month. For example, a JUN 44 is an option with a 44 strike expiring in June. Expiration is the third Saturday of the expiration month, and the *last trading day* is the Friday preceding it.

An *underlying* is the stock or other security on which the option is written. This cannot be changed. For example, a CVX Jun 44 call is a call on Chevron (CVX) expiring in June and with a strike of 44, or $44 per share.

An option's *premium* is the current value of the option. For example, an option with a premium of 1.14 is currently worth $114 for contract; for $114, a trader controls 100 shares of the underlying.

The value of each option is determined by time and proximity. The time remaining until expiration is important because time value declines as expiration approaches. Proximity refers to how close the current price per share is to the strike of the option. An option that is *in the money* will have more value because of its proximity (the underlying price is higher than a call's strike or lower than a put's strike). An option is *out of the money* when the underlying price is lower than the call's strike or higher than the put's strike. When the underlying price and strike are the same, the option is *at the money.* The status of an option relative to the underlying price is referred to as its *moneyness.*

Tip: To better understand why options prices change, study them in the context of proximity between strike and underlying price, also known as the option's moneyness at the moment.

Whenever the moneyness exceeds five points above or below a strike, the condition is called *deep in the money* (underlying price five points or more above a calls' strike or puts five points or more below a put's strike) or *deep out of the money* (underlying price five points or more below a call's strike or five points or more above a put's strike). Moneyness is summarized in figure 8.1.

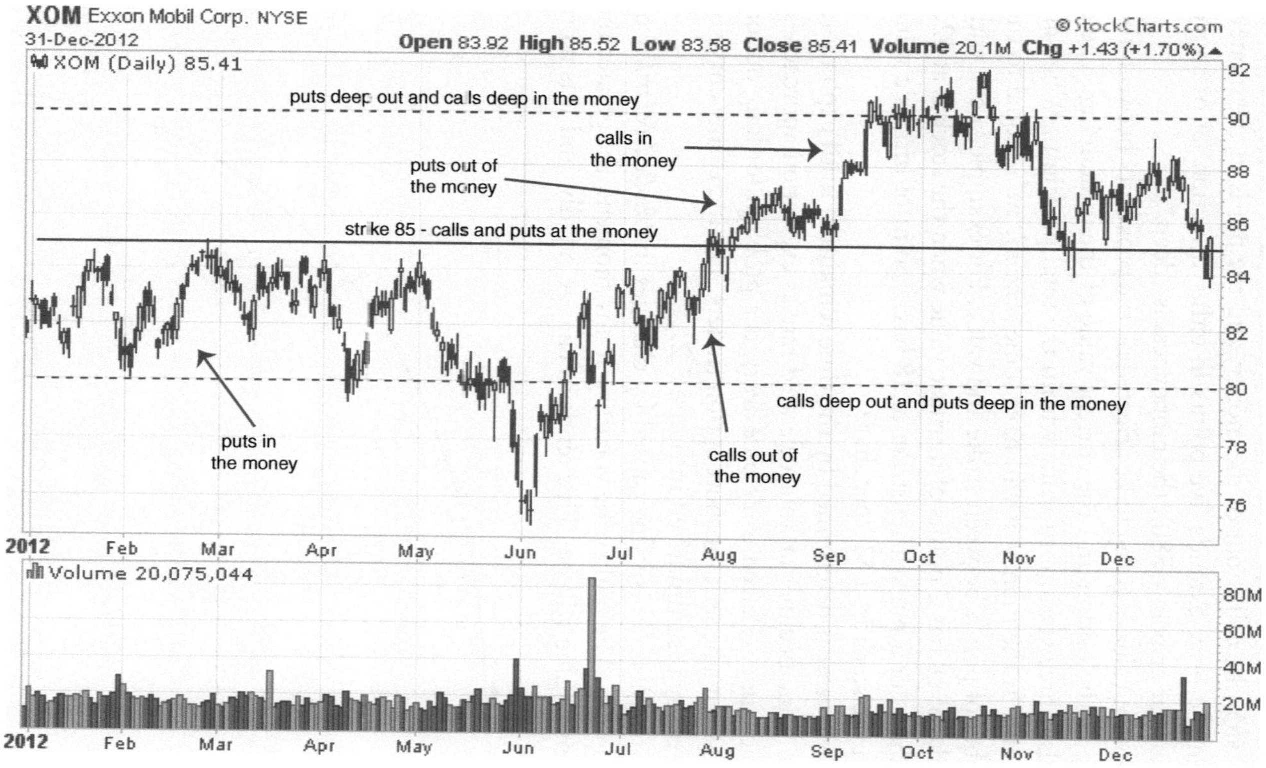

Figure 8.1 Moneyness of options

In this example, the strike of the options is 85 ($85 per share). When the price of the underlying is higher than $85, the call is in the money, and the put is out of the money. When the price is lower than $85, the call is out of the money, and the put is in the money. When the price extends five points in either direction (at or above $90 and at or below $80), the moneyness is "deep."

Option value is going to be affected by the proximity between the strike of the option and the current value of the underlying. There are also three versions of value within the total premium. When an option is in the money, it will have some *intrinsic value* equal to the number of points between strike and value. For example, a call with a strike of 18 has 2 points of intrinsic value when the stock is at $20 per share, or a put with a strike of 18 has 1 point of intrinsic value when the stock is at $17 per share.

Time value is determined by the time remaining until expiration. Time value declines at an accelerated rate as expiration approaches. This change, referred to as *time decay*, is a problem for long option traders but a benefit for short option traders. The increasing curve of time decay is summarized in figure 8.2.

Extrinsic value is value caused by changes in *implied volatility* of the option. The higher the volatility at any moment, the greater the extrinsic value will be. This tends to change rapidly, and it is more

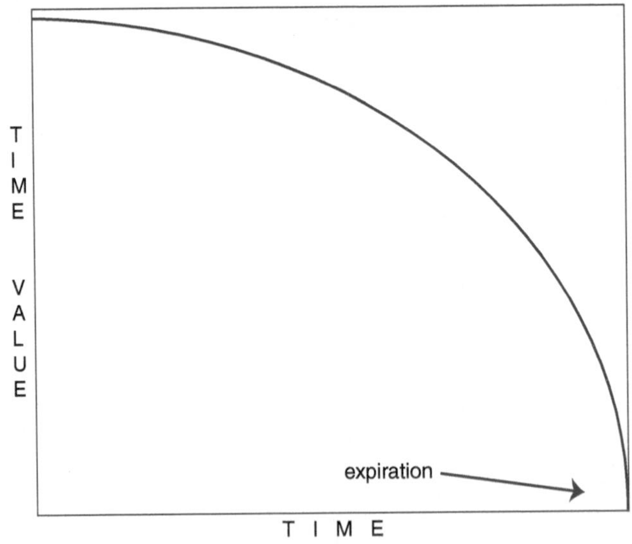

Figure 8.2 Time decay

active when the option is at the money or in the money and also when expiration is close. Long-term options tend to be very unresponsive to movement in the underlying because the time factor is not as important as it will be for options expiring in the next one to two months.

THE BASICS OF OPTION TRADING

The value of every option relies on two primary attributes: time and proximity.

Time remaining until expiration is one of the two factors adding to value. The longer the time, the higher the time value. Because time decay accelerates as expiration approaches, it is difficult for a long option to become profitable even if the price is moving in the desired direction. As intrinsic value grows, time value declines and may easily outpace growth of the intrinsic value. For example, you purchase a call with a 45 strike and pay 1.92 ($192). As expiration approaches, the underlying price begins to rise and approaches the level of $45 per share; however, the option's value does not rise. The stock has gone up three points during the last month, and the option premium has only grown by 0.50. Why? This occurs because time decay is accelerated during the final month. The underlying may even move in the money, but time decay erodes the overall value so that the option may not become profitable.

Key point: Time decay accelerates as expiration approaches and is most rapid during the last month of the option's life. This provides clues to the timing of both long and short option trades.

This problem for the owner of a long option is an advantage for the option seller. When you sell an option, you receive payment, and you hope that the value will decline so it can be closed for less money, creating a profit. Given the same facts as above, if you *sell* that call, you receive $192. As expiration approaches, the value declines. As long as the underlying is worth less than the strike ($45 per share), the entire option premium is nonintrinsic. By the expiration date, that option will be worthless. That is, if two or three days before expiration the underlying is at $44.50 per share, the 45 call may be worth only 0.25. You can enter a buy to close order, pay $25 and close the position. This means the transaction ended up with a profit of $167 ($192—$25) before trading costs.

Time is a problem for long option traders, but an advantage for sellers.

Proximity is the second feature in the valuation of options. Any option out of the money is going to be less responsive to price movement of the underlying than at-the-money positions, and as an option moves in-the-money, the creation of intrinsic value makes that option most likely to track underlying price movement. For this reason, short-term traders prefer to buy at-the-money options. The price does not include a lot of time value, and if the position moves in the money, profits are likely to accumulate rapidly. However, sellers will benefit the most when selling options expiring in one to two months, preferably at the money or slightly out of the money. These options have no intrinsic value, and time value will decline rapidly during these last few weeks, especially during the last month before expiration. Because a seller begins the transaction by receiving cash for the short trade, the rapid decline in time value provides the greatest likelihood for profits. Even if the option moves in the money, the decline in time value makes it likely that the position can be closed at a net profit.

Time and proximity work together to define value. Added to these two factors, volatility (the third form of premium, called implied volatility or extrinsic value) reflects the level of risk in the option. The greater the volatility in the underlying stock and the higher the volume of *open interest* (the current active option positions) are, the higher implied volatility tends to be. However, implied volatility also becomes highly unreliable during the final month of the option's life. The phenomenon known as *volatility collapse* changes everything. You cannot rely on the normal causes of implied volatility to affect prices, and this means you have to rely on underlying price movement and chart patterns to time trades.

Tip: Remember that in the final weeks of the option's life, volatility collapse makes it more difficult than ever to judge option values based on the underlying stock. You need to rely on proximity (moneyness) to time entry and exit.

Most explanations of options value lump implied value in with time value, but this is a mistake. Time value is highly predictable; it declines slowly when many months remain in the option's life and accelerates as expiration approaches. Implied volatility is going to change based on volatility in the underlying and on news or rumors (earnings surprises, mergers and acquisitions, or changes in guidance—estimates of future revenue and earnings—for example).

Risks and Rewards

Options can carry high risk or be highly conservative, all depending on the strategies you use. This makes the options market exceptional because risk and reward levels are determined by traders and their strategies rather than by market conditions. Options may be profitable in rising or falling markets and even in markets that have paused and do not move much at all. Because the determining factor is volatility of the underlying stock (rather than the entire market), a stock-based option and strategy should be selected based on the fundamental and technical attributes of the company and not based on the overall market or its conditions.

In the energy sector, the fundamentals include the financial status of a specific company, which affects the stock price; larger market fundamentals also affect stock prices. These include political, economic, and other factors (including perceived image of a company). For example, a company like Exxon Mobil, which is criticized by environmental interests, is not necessarily taking any actions that harm the environment more than other companies, but the perception is more important than the reality of the matter. In the energy sector more than any other, perception is a key fundamental and may easily affect market value and risk of all companies in the sector. Likewise, the perception that the planet is running out of fossil fuels is contradicted by discoveries of vast reserves (especially of natural gas) and the United States holds vast but untapped reserves of many types of fuel. However, the perception that fossil fuels are about to peak and become depleted, while false, is more important than the facts. The *perception* factor in the energy sector is as important as profit and loss.

On the technical side, the energy market is perceived risky because many companies exhibit volatility in stock prices. Stock prices move every day based not only on the rather limited influences of actual supply and demand, but also due to geopolitical and cultural disputes. The belief, widely held, that fossil fuels are the cause of global warming has not been clearly established as fact, but the belief itself affects prices of energy stocks and futures. This is seen in the charts of oil companies. The Exxon Mobil chart shown earlier in this chapter makes this point; the breadth of trading during the year shown on the chart was usually about five points over a weekly period and as high as ten points at times.

> **Tip:** The most effective measurement of price volatility is the breadth of trading. The larger the breadth, the higher the volatility. Even stocks moving quickly from high to low may have relatively low volatility.

This volatility can be turned into an advantage for options traders, who may use options to hedge risk or to make short-term profits. Higher volatility means greater risk, and reducing this risk is best accomplished through the use of options, but short-term trading may also benefit from rapid price movement. The difficulty in this is that while profits may materialize rapidly, losses have an equal chance in high-volatility stocks.

Selling Options

Selling options is among the most appealing strategies, for several reasons. First, the money flows to the seller, whereas the buyer has to pay to open a position. Second, time works to the seller's advantage, but is a disadvantage to the buyer.

For the following discussions, three options listings on energy stocks are provided in tables 8.1, 8.2, 8.3, and 8.4. All of these were current prices as of on May 6, 2013.

This is a relatively low-priced stock, so the option values are also low. However, this is not a valid reason for rejecting this as a potential options play. A stock selling at $18.86 with a call valued at 0.82 is the same as a stock selling at $94.30 with an option at 4.10.

The lower-strike calls (17 and 18) are in the money, so they tend to be worth more premium than the 19 strikes that are out of the

Table 8.1 Petro Brasileiro S.A. options listings

6-May-13	STOCK	MONTH	STRIKE	CALLS	PUTS
	$18.86	JUN (47 days)	17	2.17	0.27
			18	1.41	0.51
			19	0.82	0.94
	$18.86	JUL (75 days)	17	2.33	0.41
			18	1.62	0.70
			19	1.03	1.14

Table 8.2 British Petroleum options listings

6-May-13	STOCK	MONTH	STRIKE	CALLS	PUTS
	$44.03	JUN (47 days)	43	1.34	0.84
			44	0.78	1.29
			45	0.40	1.92
	$44.03	JUL (75 days)	43	1.64	1.14
			44	1.09	1.60
			45	0.68	2.21

Table 8.3 ConocoPhillips options listings

6-May-13	STOCK	MONTH	STRIKE	CALLS	PUTS
	$62.10	JUN (47 days)	60	2.44	0.88
			62.5	0.88	1.96
			65	0.23	3.70
	$62.10	AUG (103 days)	60	2.82	1.87
			62.5	1.43	3.15
			65	0.62	5.10

Table 8.4 Chevron options listings

6-May-13	STOCK	MONTH	STRIKE	CALLS	PUTS
	$123.39	JUN (47 days)	120	4.20	1.69
			125	1.33	4.00
			130	0.25	7.95
	$123.39	JUL (75 days)	120	4.85	2.41
			125	2.06	4.70
			130	0.64	8.30

money. The opposite applies to puts: The 19 strike is in the money, and the 17 and 18 strikes are out of the money. The values move in opposite directions because calls gain as the underlying price rises, but puts gain as the underlying price falls.

The BP listing of options, like the previous one, is based on one dollar strike increments. This is normally the case for stocks under $50, but exceptions can also be found.

The COP listings have 2.50 increments, typical of stocks priced between $50 and $100. Many of these are reported in increments of 5.00, however. In this listing, there were no July options available as of May 6 (although a set of July would show up in the month of June; at a minimum, options are available for the current month and the month following and then every three months based on the option cycle for the company.

Chevron reported options with five dollar increments. The stock was selling at over $100. In this price range, you may find some companies with 10.00 increments but few with increments under 5.00.

The risk in short selling is that when the underlying moves in the wrong direction (up for short calls or down for short puts), losses can and do occur. And the farther the price moves, the bigger the loss. An *uncovered option* (also called a *naked option*) is a short position that is exposed to the possibility of loss. If you sell a short call or put, you have to worry about the potential loss you may experience as a consequence.

> **Tip:** Although naked options are said to have "unlimited" risk, the market places limits on a reasonable maximum movement an underlying may experience over time.

For example, you sell one CVX July 125 call at 2.06 and receive $206. The stock was priced at $123.39 at the time of your sale. So the short call was entirely out of the money. There are 75 days to go until expiration, making this ideal option in terms of timing. You should expect time value to decline quickly over this period of time. The risk is that if the stock price rises above the $125 level, the short call will move in the money. You risk that the call will be exercised. In that case, you could experience a loss. The farther in the money it moves, the bigger your loss. The actual breakeven is $127.06 ($125 strike plus 2.06 received for selling the call). If the stock moves higher than that, you will suffer a loss.

Exercise can be avoided by rolling forward. This involves buying to close the current short option and selling one that expires at a later date at the same strike or at a higher strike. If you sell at the same strike, you will earn a net credit; if you sell at a higher strike, a credit is possible but a small debit is more likely. In theory, you can roll forward indefinitely, and exercise is most likely right before expiration. However, *early exercise* is possible, especially right before ex-dividend date. If your short call is in the money before this date, it could be exercised by another trader who wants the current dividend. So if you write an uncovered call, you need to be aware of when ex-dividend date occurs and of the danger of early exercise.

In theory, an uncovered call has unlimited risk because the stock price could rise indefinitely. In practice, it will only rise to a degree based on historical price range, P/E ratio, and other mitigating fundamentals. In comparison, an uncovered put has much more limited risks because the stock price can only fall to zero. And realistically, the true risk is not zero, but the tangible book value per share (book value minus intangible assets).

For example, you sell an uncovered BP June 45 put and received 1.92 ($192). At the time, the stock price was $44.03. So this put was 0.97 in the money. The option expires in 47 days, so you can expect a rapid decline in time value, and of the premium of $192, $95 is time value. This is a good example of the seller's advantage: Even if the stock price remains in the money, the premium will decline by $95 over the next 47 days, so the options' value would fall to $97, and the position could be closed and a $95 profit taken. The danger, however, is that if the stock price falls, the option would gain value, meaning that as a seller you would incur a loss. Your breakeven on this was $43.08 per share ($45 strike minus $1.92 received for the put). If the price of the underlying falls below this level, you will incur a loss. However, to avoid the loss, the short put can be rolled forward—closed and replaced—with a later-expiring position at the same strike or at a lower strike.

Tip: Calculating net breakeven (basis in the underlying adjusted by the value of the option) is crucial to identifying levels of risk and to knowing when to exit a trade.

In addition to the market risk of uncovered options, another point to keep in mind is how margin works for options. Margin for stock purchases is a leveraging device. You can borrow up to 50% of your purchase price, which means that you can buy 100 shares at $50 per share with $2,500 cash and $2,500 borrowed on margin. In the options version, margin is not leverage but a specific requirement for collateral, or placing cash and securities in the margin account to cover risks.

Margin for options refers to collateral requirements. You are required to maintain adequate collateral based on the nature of the trade. An uncovered option comes with a margin requirement equal to 100% of the exercise value. So an uncovered option with a 50 strike requires margin collateral of $5,000. An essential reference for all options traders is the CBOE Margin Manual, which can be downloaded free at http://cboe.com/LearnCenter/workbench/pdfs/MarginManual2000.pdf.

The uncovered option is a high-risk venture. But you can also sell a *covered call,* which is a highly conservative strategy. In this strategy, you own 100 shares and sell a call. The shares "cover" your exercise risk, removing the major risk from the position. A put cannot be covered in the same manner as a call.[3]

If you open a covered call, you want to make sure that the strike is higher than your basis in the stock. In the event the call is exercised, you want it to result in a capital gain and not in a capital loss. The call premium discounts your basis in the stock, which mitigates stock-based market risk. For example, you buy 100 shares of CVX and pay $123.39 per share. You also sell a June 125 call for 1.33 and receive $133. The covered call reduces your basis to $122.06 ($123.39—$1.33). It also sets up a net gain in the event of exercise in the amount of $1.61 per share ($125.00 exercise—$123.39 per share basis). So in the covered call, you have three sources of income: capital gain on the stock, option premium, and dividends.

Selecting the best covered call is a matter of comparing contracts on an annualized basis. Calls expiring later usually yield higher dollar amounts, but they will come out to less money and a lower percentage on an *annualized basis.* The options expiring sooner are usually more profitable. For example, you bought100 shares of PBR at $18.86 and you are considering selling a covered call with a 19 strike. The June call is worth 0.82 with 47 days to expiration, and the July call is worth 1.03 with 75 days to expiration. The difference of $56 is significant, so you are tempted to go for the August call and more cash.

However, comparing these choices on an annualized basis tells a different story. You annualize by restating your return as if a position were held for exactly one year. Using the purchase price as a basis for this comparison, the two options yield initial returns of:

June: $82 ÷ $1,886 = 4.30%
July: $103÷ $1,886 = 5.46%

To annualize, the initial returns are divided by the holding period and then multiplied by a full year. The holding period may involve days, weeks, or months. In this example, the number of days is used to show how annualized return changes the comparison between these two options:

June: (4.30 ÷ 47) x 365 =33.39%
July: (5.46 ÷ 75) x 365 =26.57%

The annualized result for the June option is far better than that for the July one. This does not mean you should expect to earn more than 30% on covered call writing consistently, but it does demonstrate that the shorter-term option is more profitable. Looking at this in another way, in a comparison of 47-day to 75-day positions, you could write nearly eight covered calls per year (365 ÷ 47 = 7.77) versus only five per year using the July option (365 ÷ 75 = 4.87). On the basis of dollars you would earn under each scenario and based on the current premium levels:

$82 x 8 = $656
$103 x 5 = $515

Looking at the comparison in this manner also proves that the options expiring sooner are more profitable than those expiring later.

Options on Futures: Leverage on Leverage

One of the highest forms of leverage is trading options on energy futures. A futures contract is a form of leverage on its own, and futures options leverage the futures contract so that potentially large returns are possible. This is a very attractive method for trading in the energy sector.

An option on a futures contract is written against a single contract (in comparison, an option on stock is written on 100 shares). The buyer acquires the right to exercise and acquire that futures contract at any time before expiration. A seller of an option on a futures contract agrees to deliver that contract if and when the option is exercised.

> **Key point:** Options on futures and stock are *not* the same. As with all forms of trading, know the rules before making a move. This will save you money by helping avoid errors.

These options usually expire near the end of the month just before the delivery month of the futures contract. Thus, if the futures contract's delivery date comes up in November, the option will expire in October. A distinction has to be kept in mind: Unlike stock-based options in which the underlying is 100 shares of stock, the futures option is based on the futures contract and not directly on the commodity itself. So the futures option is leverage on leverage, since a futures contract is already a form of leverage on the underlying commodity.

Futures are highly specialized and energy investors can restrict their activities to stocks or ETFs and, in either case, also use options to control risks while generating income. Options can have a high risk or be very conservative, and a majority of investors are going to be interested in designing a conservative strategy to enhance a portfolio of equity positions in the energy sector.

A RANGE OF STRATEGIES: FROM CONSERVATIVE TO HIGH RISK

Conservative options strategies energy traders can use include the basic covered call, collar, variable ratio write, and synthetic stock positions. High-risk strategies include writing uncovered options and leveraged day trading strategies.

Covered calls are the favorite strategy among conservative investors. This strategy consists of buying 100 shares of stock and then selling a call. In the event the stock rises and the call is exercised, the shares are called away at the fixed strike. If the stock price remains at or below the strike, the call will expire worthless. Alternatively, it can be closed with a "buy to close" order and profits taken. Once the call is closed, a new covered call can be opened. The appeal to this strategy is that you continue earning dividends while the call is open.

A danger in covered calls is that the stock price may decline below your net basis. The net basis is your purchase price of shares, reduced by the premium received for selling the option. For example, you buy 100 shares of BP at $44.03 and sell a July 45 call for 0.68 ($68). Your net basis is $43.35 ($44.03—$0.68). In this event, you need to wait for the stock to rebound, sell at a loss, or write a subsequent covered call. However, you do not want to write any call that will create a net loss.

Collars are expansions of the covered call. In addition to selling a call, you also buy a put. This protects you against downside price risk. If the price does decline, the put can be closed at a profit, with gains offsetting the loss in stock. Or the put can be exercised, so that 100 shares will be sold at the fixed strike although the current price per share is lower. For example, you buy 100 shares of PBR at $18.86 per share, sell a July 19 call for 1.03, and buy a July 18 put for 0.70. In this situation, you have a credit on the two options of $33 ($103—$70). You have protected the position against downside movement below the put's strike. However, you have also limited potential profits to a maximum of $47 ($33 from the options plus $14 in capital gains if the call is exercised).

There are situations like this where you are willing to accept a very limited profit in exchange for protection against price decline. PBR yields a dividend of 6.3%, so if getting the dividend without risk is your primary purpose, then a collar makes sense. You remove all market risk in the stock below the put's strike in exchange for a limited potential profit in the stock, but you earn the quarterly dividend. For many traders, this is preferable to simply purchasing shares of stock and earning the dividend but having to accept the risk of price decline.

Variable ratio writes are another expansion of the covered call. In the basic covered call, you sell one option for every 100 shares you own. A ratio write involves selling more options than you cover. For example, if you own 300 shares and sell four calls, you create a 4:3 ratio. This is a combination of three covered calls and one uncovered call. The idea is that for the additional premium, the added risk is acceptable. However, if the stock moves high and all of the options move in the money, the ratio write could become quite risky. A way to reduce this risk is with the *variable* ratio write. In this version, you employ two strikes.

For example, you buy 300 shares of COP at $62.10 per share. You also sell two August 62.50 calls at 1.43 and two August 54 calls at 0.62. This generates a total income of $410. All of the options are out of the money. If the stock price moves up beyond the lower strike of $62.50, one or more of the 65 calls can be closed to avoid losses resulting from exercise. Or any or the calls can be rolled forward. The variable ratio write reduces risks because the two different strikes are employed, giving you control over the risks while profiting from shorting the calls.

Synthetic stock is a strategy employing a single strike and opening two options at the same time. It is called "synthetic" because the option values move exactly like the value of 100 shares of stock, but the cost of the two positions is at or close to zero. In a synthetic *long* stock position, you buy one call and sell one put. The danger here is that if the underlying price declines, the short put gains in value and could result in a loss. However, the short option position can be closed or rolled to avoid exercise, a step you cannot take if you hold 100 shares of stock and the price falls. If the price rises, the long call gains in value and tracks the underlying point for point. In a synthetic *short* stock position, you buy one put and sell one call. If the stock price rises, the short call gains value and may lead to losses; however, it can be closed or rolled to avoid the loss. If the stock price declines, the long put gains one point of value for each point the stock loses.

For example, you open a synthetic long stock position on PBR by buying a June 18 call at 0.82 and selling a June 19 put at 0.94. Your net credit on this transaction is $12. If the stock rises above $19 per share, the short put becomes worthless, and the long call tracks the stock point for point. It synthetically mirrors the price movement. If the stock price declines, the short put becomes more valuable, which means that you lose point for point. However, you can also close or roll the short put to avoid that point-for-point loss to the downside. By reversing these two positions you can create a bearish synthetic stock position, which gains value if the underlying price drops. This involves buying the put and selling the call, setting up a $12 debit. Your synthetic position gains one point for each point the stock declines, and you lose one point for each point the stock rises.

PROFITABILITY WITH OPTIONS

As an investor in the energy sector, you can build respectable profits by combining positions in stock with options and dividends. Many strategies are designed to ensure consistent profits. In the energy market, you can use strategies such as covered call writing to create current cash income based on stock ownership without adding market risk to the stock position.

> **Tip:** When calculating net return, be sure you apply the same consistent standards to each case so your comparisons are valid and accurate.

For example, a portfolio of energy stocks can help you to set up your own portfolio that acts like an ETF, and it also allows you to gain dividends and option benefits. You could create a $50,000 portfolio of the four energy stocks previously described, write covered calls, and collect dividends on all four (see tables 8.1–8.4 for illustration of the following):

PBR—Petro Brasileiro S.A. 6.3% dividend
  buy 600 shares @ $18.86 = $11,316
    sell 6 covered calls, June 19 @ 0.82 = $492

BP—British Petroleum 4.8% dividend
buy 300 shares @ $44.03 = $13,209
sell 3 covered calls, June 45 @ 0.40 = $120

COP—ConocoPhillips 4.3% dividend
buy 200 shares @ 62.10 = $12,420
sell 2 covered calls, June 62.50 @ 0.88 = $176

CVX—Chevron 3.2% dividend
buy 100 shares @ 123.39 = $12,339
sell 1 covered call, June 125 @ 1.33 = $266

Total portfolio
Stock value = $49,284
Option premium = $1,054
Base option return ($1,054 ÷ $49,284) = 2.14%
Annualized option return (2.14% ÷ 47 x 365 = 16.62%
Average dividend yield = 4.65%
Total annualized return (16.62% + 4.65%) = 21.27%

The purpose of annualizing is to demonstrate the combined return of options with dividends. However, this analysis assumes that none of the short calls will be exercised (in the event of exercise, small capital gains will be realized as well, and dividends might not be earned as a consequence). This makes the point that a portfolio of high-yielding energy stocks combined with covered call writing can be very profitable.

Is this adequate diversification? It is not a diversified portfolio because all of the holdings are energy stocks. Three points are worth making about this: First, if you purchase shares in an energy ETF instead of holding stock, you have an equally undiversified position. Second, with an ETF you may not have as much control over options positions, and you probably will not earn as high a yield on dividends. Third, if you want to modify a portfolio to diversify beyond energy, you could combine one energy stock with stocks in other sectors. In this manner, you continue earning high dividends as long as you continue to focus on high-yielding stocks, and you still write covered calls to achieve double-digit annualized returns.

Options help you to leverage your portfolio. However, this market demands skill and experience and the jargon of options trading is daunting. If you want to use options to leverage energy stocks, you should first learn how options work and study the risks of the many possible strategies that you can use to manage your portfolio.

Part III

Investment and Trading Strategies

CHAPTER 9

KNOWING YOUR RISK TOLERANCE

> All this worldly wisdom was once the unamiable heresy of some wise man.
>
> Henry David Thoreau, *Journal*, 1853

RISK. IT'S THE OPPOSITE SIDE OF OPPORTUNITY, BUT THAT IS OFTEN not acknowledged by investors.

Risk is a poorly understood aspect of investing. Several tendencies and assumptions cause this, including the following:

1. *Zero-point entry assumption.* The most basic error investors make is to assume that their entry price is the zero point and that from there, prices will rise. In fact, any price is only the latest, and it can rise or fall. Understanding this feature alone will help avoid most surprises. Everyone has had the experience of seeing prices fall right after buying shares; you just have to understand that prices rise *and* fall. This seems rudimentary, but so many investors operate on the flawed assumption that it is worth reminding yourself of this fact.

Key point: Price movement is continuous; your entry point is not *your* zero price point, but the latest price in a series that may continue either upward or downward.

2. *Refusal to cut losses and tendency to double down.* A wise chess player who loses a piece knows to go on the defensive; a less

experienced player wants to get aggressive to even the score and becomes reckless, usually losing. The same tactical observation applies in investing. It makes sense to close a losing position as soon as you see that it is not working so that you can avoid further losses. Learn from the mistake and move on to the next position. It's a big mistake to accept greater risks by doubling down to get back what you have lost.

Key point: The expectation of winning in every trade is dangerous and ill-advised. You can expect to use analytical skills to improve your wins, but you also need to be able to walk away from losses—without looking back.

3. *Basic poor understanding of fundamentals.* The fundamentals include financial status and profits of a specific company or market-wide economic trends. Some technical traders discount the importance of fundamentals because they are outdated, uninteresting, or complex. But you can reduce your list of investment candidates with a short list of indicators that measure financial strength or weakness, cash flow, and profitability. Even a technical trader will gain from applying a few of the fundamentals to the selection of companies or products.
4. *Unawareness of how to diversify risks.* To some, diversification means buying shares of several different companies. However, if all of those companies are vulnerable to the same economic and market forces, the portfolio is not diversified at all. Effective diversification is not based on holding several different positions, but on diversifying *risk exposure* among elements of the portfolio.
5. *Refusal to acknowledge all forms of risk.* Most investors know all about market risk, which concerns price changes. The risk that value will fall instead of rise is for many the only true risk in investing. But in fact risk comes in many different forms, and market risk is only the starting point in determining how to build a portfolio that fits with your risk tolerance.
6. *Focus on profit potential only.* In evaluating a company and deciding whether to buy its stock or a commodity to invest

in its futures or options, it is logical to understand the profit potential each decision involves. Equally logical is an evaluation of the loss potential, or risk, accompanying any decision. The tendency is to focus on best-case outcomes and to overlook the worst case at the same time. By evaluating the full range of possible outcomes, investors improve their decisions and, ultimately, their profits.

Key point: Every best-case outcome is offset by an equally possible worst case—and to know about both is common sense and the most effective way to manage risk.

THE NATURE OF RISK

Energy investors and traders face higher than average risks due to the high volatility of this sector. Risk comes in several forms, and as an energy investor, you need to be aware not only of market risk but also of the full range of additional risks that exist.

The best known risk is market risk, and it is more accurately called volatility. The danger of falling prices is offset by the potential for rising prices. In this regard, risk and opportunity are different symptoms of the same exposure in the energy sector and elsewhere. Market risk, however, is more than just the risk of unfavorable price movement. Potential losses have to be quantified in order to understand degrees of risk, and this is no easy task. On the surface, it would seem that some obvious measurements can be applied; for example, you might conclude that a $40 stock is twice as risky as a $20 stock because it could fall 40 points or twice the maximum loss in the lower-priced stock. But this would be inaccurate. To determine the true market risk of the two stock investments you have to take into account the likelihood of risk occurring at its maximum level. The $40 stock might represent a safer investment than a lower-priced stock based on its fundamentals. In that case, it contains less market risk, not more.

The application of *Value at Risk* (VaR) helps quantify risk. This measures the risk of loss in a portfolio. For example, if a portfolio is exposed to a one-day 5% VaR, this means there is a 5% chance that the portfolio value could decline by 5% if no trading occurs. This

5% threshold is important, and anything higher than a predetermined level is termed a VaR break.[1]

The assumption used in VaR that no trading occurs on the day of a probability test, is that losses can be observed in that situation, whereas trading transactions distort outcomes and make them more difficult to measure. The formula is intended to identify the level of maximum likely loss on any one day or in another timeframe.

VaR is not necessarily a positive force in risk management. Although risk managers and portfolio managers may rely on VaR, it can also be easily misunderstood or misused. VaR can provide a false sense of security, convincing investors that losses are not possible when they are. This was cited as one of the underlying problems in the financial meltdown of 2007–08.[2] One reason for this problem is that when dealing with an extensive portfolio of many equities, futures, or derivatives, the overall VaR might be much higher than the individual VaR for each component in that portfolio.[3]

Like most models, VaR is flawed in the sense that it assumes a portfolio remains unchanged over a long period of time, which is rarely the case. Over extended time periods, position values are going to change, and the longer the period, the less reliable VaR becomes. On a practical level, the complexity and potential unreliability of VaR mean that it may not be used exclusively in quantifying risks of portfolio positions. While VaR may work well if you assume today's values apply in the future (and thus are "normal"), it simply provides a threshold value that is limited by both probability and time.[4]

Tip: Risk modeling with the use of VaR and similar tests may help quantify risk levels, but it will not provide as solid a test of risk as old-fashioned analysis of fundamentals.

For a majority of investors and traders buying shares of stock, futures, or options in the energy sector, VaR may not play as great a role as more fundamental tests of competitive position, working capital, or long-term profitability trends. However, the statistical modeling derived from VaR makes the point that risk is not merely a measurement of how many dollars an investment may lose in the worst-case outcome, but a quantification of risk itself. For example, when deciding whether to buy shares, you are applying a set of assumptions to determine whether today's price is reasonable and what the chances are of the investment growing in value or losing value. Most important of all, you must determine when you will take profits or limit losses and close the position.

Even long-term buy-and-hold investors need to set goals and limits for themselves, performing without statistics what VaR proposes to do for a larger portfolio. So you set your probabilities, and, of course, you would not take up a position unless you believed there was a reasonable chance of profits. Likewise, every investor creates a set of probabilities, even though the precise level is not always stated. However, the exercise of trying to set up probabilities is valuable in identifying risk and matching it to risk tolerance.

As a starting point, few investors would be comfortable placing money into an investment with a fifty-fifty chance of profits. These are not great odds; however, without *any* risk analysis, you might be making a series of fifty-fifty bets. (Note the use of the term "bets" in place of "investments," implying that without risk analysis, the decision to take a position is a bet.)

Key point: Know the difference between a risk and a bet. Your purpose may be to identify this difference as smart risk versus simple luck.

How do you identify the risk so that a fifty-fifty is changed to a level at which you will place an order? For most investors, reliance on fundamental, technical, or combined analysis is a method for developing a likely portfolio or compiling a list of candidates that may or may not fit. In the previous chapter, a sample of this was provided based on the assumption that you would want to create a portfolio of energy stocks.

The first issue to consider is that of diversification. An all-energy portfolio might contain too much risk in terms of lacking diversification. The same portfolio consisted of high-yielding dividends combined with covered calls. Given the market risk of covered call writing (the risk that the stock value may fall below the net investment value), a prudent approach would be to combine energy stocks with other, nonenergy stocks also yielding higher than average dividends. However, if the purpose is to create the same effect as an

ETF but with greater control, then the question of diversification might not even matter. This question is addressed in more detail below, but for the moment, as a means for quantifying your risks, the question comes down to this: Does creation of a portfolio containing specific characteristics (market cap, profitability, debt ratio trend, PE, dividends, options) mitigate risks adequately? If you do not apply any of these tests and just pick energy stocks because you think they are going to increase in value, then you make a decision to build a portfolio with nonquantified risks. It may be preferable to set up a random fifty-fifty portfolio rather than to accept one with *unknown* levels of risk. A fifty-fifty risk portfolio has an even chance of becoming profitable, but a portfolio picked haphazardly or for misguided assumptions may have less chance than the fifty-fifty one.

> **Tip:** If you just want to take your chances with the market, random selection might achieve a fifty-fifty exposure to profit. In comparison, performing only a few basic tests may vastly improve your chances, so there are good reasons to test profitability, P/E, and dividends.

In addition to diversifying, you quantify the risks of your portfolio through identifying supplemental means of income generation. If you focus only on a buy-and-hold strategy, you may select stocks solely because of low volatility and a long record of modest or consistent profits. You may be uninterested in supplemental means of income generation, including dividend yield or options premium. In that case, the process of picking extremely conservative stocks is considered safe, but it might not match or beat inflation or, worse, the combined risk of inflation and taxes. So if you are to quantify risk, risk analysis has to be performed realistically. You can apply the following guidelines:

1. *50/50 is not that good.* The realization that a fifty-fifty portfolio has an equal chance of profits or losses may be reassuring to some. That is potentially a better likelihood of profits than many professionally managed mutual funds, which consistently underperform the market. However, while fifty-fifty is

preferable to poor selection, it is not good enough as a standard for investment performance. By applying fundamental criteria, you should be able to eliminate products whose likely profits are dismal, favoring those with greater potential. This increases fifty-fifty to a level that, while impossible to assign a percentage to, is going to be better than fifty-fifty.

2. *Random selection is not much better.* Some investors, even those calling themselves fundamental or conservative, make investment decisions in a random manner. They hear a news story about a company or read an article or hear friends talking about their investments, and then they decide to buy shares. If someone tells you to buy shares of a company, ask a few questions, such as: What business is the company in? What were last year's profits? What is the company's dividend yield? You will discover that most people giving you stock tips cannot answer these questions because they haven't done their homework. If you invest on the basis of their advice, you're not doing your homework either. As they say in school, don't copy answers from the person at the next desk if he or she didn't study either.
3. *Supplemental income (dividends and options, for example) are significant in reducing risks.* You might be like most investors, interested in the energy sector for all of the right reasons but not sure how to reduce market risks. Looking for *conservative* ways to reduce risk makes a lot of difference and may move an investment of unknown quality into the camp of high quality. For example, high-yielding dividend stocks are going to yield more than those of a comparable company in the same business paying little or no dividend. Likewise, if you write a series of covered calls, you still face market risk, but that risk is lower than just owning shares because the premium you earn reduces your basis in stock. It generates a stream of income that, when annualized, may take you into double digits. For example, the portfolio described in the previous chapter produced 16.62% annualized return from covered calls based on a holding period of 47 days and that without even counting dividends. Options selected with expirations in one to two months will almost always annualize out to double-digit returns.

Key point: In the overall risk evaluation of a particular stock, reducing your list to stocks yielding high dividends and adding in option income will vastly reduce your market risk.

4. *So-called safe investments (in terms of low volatility) may be unacceptable because they do not match inflation, leading to a loss of value in your capital.* The effects of inflation on your portfolio value cannot be overlooked, but a majority of investors do not consider this. For example, if inflation is typically 3% per year, a portfolio yielding 2% is losing purchasing power. That inflation rate means your capital value loses 3% per year. For example, every $100 of invested capital declines to $97 when inflation is 3%. If your portfolio earns only 2%, you are not keeping up with inflation. This presents a scenario of invisible market risk, which can be defined as the risk that your portfolio is profitable but not profitable enough to match or beat inflation. That is a real loss and should not be overlooked in the analysis of what constitutes risk. If you lose purchasing power due to the net difference between inflation and portfolio gains, then market risk has to be defined as the net profit *after inflation* and not just as a percentage of return on invested capital.

BEYOND MARKET RISK

Although market risk is troubling by itself, you face an array of other risks as well. These include the following:

1. *Knowledge and experience risk.* The most frequently overlooked substantial risk is that of knowledge and experience. Investors tend to buy stock with little or no in-depth analysis; instead, they either rely on tips or want someone to tell them what to buy. A tip is rarely worthwhile; if you hear a tip, you should ask who the source is and why the tip is being given. Ask questions, basic questions that address the quality of the company and its value. Anyone who does not know how the market works should either avoid it altogether or rely on someone else for help. This is where financial professionals come into the

picture; they are well suited to help those lacking the knowledge and experience to make informed decisions.

Experience is just as important. If you are interested in futures or options, by definition you should have the experience in those markets to know how much risk you can afford to take and which strategies are appropriate. If you rely on someone else to advise you, then these markets are just not appropriate for you based on your experience. If you are attracted to stocks but you don't know how to pick appropriate companies, either rely on a financial professional or use a mutual fund or ETF to create diversification. The complexity of the energy market makes it tough to quantify risks for any individual company or strategy, not to mention deciding which energy subsector to pursue. The ETF route (or for commodities, an index fund) makes sense for anyone who does not have a sound and complete investing background.

2. *Underdiversification AND overdiversification risk.* Most people understand underdiversification, the concept of too many eggs in too few baskets. This is the most common problem that investors face in trying to manage their portfolios. Diversification occurs on several levels: within stocks and sectors, among direct ownership and nondirect (mutual funds, ETFs, index funds), or on an allocation basis (diversifying between stocks, real estate, options, and futures, for example).

   Overdiversification is the opposite problem, spreading capital among so many different products or markets that overall performance is below market average. This is a problem the mutual fund industry has created for itself. Because funds limit the percentage they can own in any one company, the bigger they get, the more expansive their diversification. Consequently, the megafunds have to be invested in the entire market; they are overdiversified. Making matters worse, management may decide to buy and sell at the worst times so that their end-of-quarter- results look as positive as possible. This might be good for management but not necessarily for the shareholders.

   Individuals can also over-diversify. Trying to eliminate all risks by investing in too many different sectors creates an overall bland average return. This reduces potential losses but profits suffer as well. One popular idea is to diversify by buying shares in several different ETFs. For example, you might hold shares in energy, retail, technology, and health care. But because each of those four ETFs are limited to a basket of stocks in those sectors, you end up with four separate undiversified holdings. If any of those sectors declines, the entire basket of ETF holdings is likely to decline as well; at best, the return will represent the average of stocks. For example, you might have 12 stocks in the energy sector, and four of them may have spectacular results while four others are typical of the overall market and four underperform. The average of all 12 will approximate the overall market, so overdiversification in this case was not a profitable decision.

Key point: It is a mistake to assume that professional management of a mutual fund will be able to outperform the market. History shows that this is not true.

The solution may be to prefer underdiversification. In other words, if you pick individual stocks in strong industries and focus on these, you are more likely to outperform the market. This is especially true when you combine stock ownership in high-yielding companies and covered calls or other options strategies that hedge market risk. Options are especially effective in risk hedging, and if you can control market risk, you do not need as much diversification as you do without that hedge.

3. *Leverage risk.* Stock investors are attracted to margin investing, because you can double your profits. Buying 100 shares of a $50 stock costs only $2,500 on margin, and the other $2,500 is borrowed. So each point the stock price rises, you gain two points in profit. But too often, investors overlook the reality: For each point the stock loses, you lose two dollars. Margin—leverage—works to accelerate not only profits, but losses as well.

   Most people would never entertain the idea of borrowing against their home equity to invest, but will use margin accounts to buy stock. It's the same thing. Leverage is a form of risk, and cannot be ignored or minimized. When you leverage an otherwise conservative investment, you increase the risk level.

Investors are further confused by the fact that in the options market, "margin" has a different definition than in the stock market. It is a collateral requirement. For example, if you opened an uncovered short option, you are required to deposit 100% of the strike value in your margin account as collateral for the risk. So selling one 50 strike option comes with a margin requirement to deposit $5,000 on margin. This is not leverage; in fact, it counters the leverage you get otherwise by trading options. Don't make the mistake of assuming that the definition of "margin" for options is identical to its definition for stock trading. Long options are another form of leverage; for 5% or less of the value of 100 shares of stock, you can buy calls or puts. This enables you to control 100 shares for a fraction of the cost of 100 shares. This is a low-risk and powerful form of leverage, but there are different leverage risks involved. With options, you have to think about the time problem. As expiration approaches, time value declines on an accelerating basis. Eventually, the option expires worthless unless it can be sold at a profit.

4. *Inflation **and** tax risk (breakeven risk).* The inflation risk is well understood in the energy sector as well as among investors in general. If the rate of inflation exceeds your return on investment, you lose purchasing power. For example, with 3% CPI, an investment return of 2% is losing 1% per year. You need to earn at least 3% just to keep up with inflation.

   The problem is more severe, however, when that investment return is taxed. The true breakeven rate is the after-tax return that matches inflation. This means that most investors need to earn a much higher return than they believe. The combined federal and state taxes can erode one-third or more of the pre-tax return on investments, so both federal and state taxes have to be taken into account to calculate the breakeven.

   For example, assume a 3% rate of inflation and a 33% federal tax rate plus a 6% state tax rate. The combined taxes reduce investment income by 39%. To calculate the breakeven return after inflation and taxes, you first need to calculate your effective tax rate. Using the example of 39%, the rate of inflation is divided by the net income after taxes:

   3% ÷ (100 – 39) = 4.9%

In this example, you need to earn 4.9% on your investments to break even. The risk for every investor is twofold. First is the requirement to earn a higher rate than previously understood; the truth is, many investors have difficulty earning the true breakeven return. Second is the increased risk resulting from increasing market risks to make or surpass breakeven.

> **Tip:** To calculate your after-tax, after-inflation breakeven point (B), divide your assumed interest rate (I) by your net income after your effective tax rate (T) – B = I ÷ (100 – T)

For many conservative investors seeking a combination of favorable dividend yield and option premium makes sense because it increases yield without greatly increasing market risk. This is especially wise when companies are selected for their value and historical track record of profits, in addition to their dividends and option premium. In the energy sector, several companies meet these criteria and point the way to beating the breakeven rate without increasing market risk.

5. *Lost opportunity risk.* A final variety of risk worth mentioning is lost opportunity. This is found in several forms. Among the most important is the equity portfolio version of lost opportunity. If you dispose of holdings when they become profitable, you may easily end up with a portfolio full of stocks whose current market value is lower than your basis—because you have taken profits when available while holding on to declining stocks. Thus, all of your capital is tied up in losing positions. The solution: If you take profits in one company, match this against a loss in another. This helps avoid lost opportunity, because the net proceeds from the sale of both is then freed up to pursue other positions—you reduce the portfolio for profits in one position while accepting losses in another. In addition, this reduces the tax burden from taking profits; the matching of profits and losses is an offset.

   Another version occurs when certain types of hedges are entered. For example, writing covered calls against shares of stock produces current income from the short call, but this also prevents you from gaining more profit if and when the stock price moves above the call's strike. For most covered call writers, this is a worthwhile risk to take. A majority of holdings will not climb so rapidly as to offset the benefits of covered calls, and in exchange, the consistent double-digit returns from options and dividends justify the lost opportunity risk.

THE RISK CAST OF CHARACTERS

No two investors are identical in the level of risk they can afford, their knowledge, and experience in trading, income, available capital, or long-term investing goals. Therefore, your energy investments should serve as part of a broader portfolio and investing strategy.

Any form of diversification has to consider the market risk as the most serious risk to the portfolio. A steep decline in value of holdings in any one sector accelerates the calamity of loss, meaning that diversification is essential as a protective measure.

Where does this leave the world of the ETF? By definition, the ETF is intentionally focused on a basket of securities with a common element. The initial appeal of this concept is that by combining a basket of securities into a single stocklike product, you create instant diversification within the sector. This is an acceptable theory as long as the entire sector does not decline in the same way or at the same time. Since energy is one of the three basic necessities (the other two being food and shelter), a disastrous loss is less likely in the energy sector than it would be elsewhere. For example, retail, food and beverage, leisure, hotel, and gaming stocks could suffer a decline due to economic conditions, with all companies in those sectors declining at the same time, possibly to a considerable depth.

> **Tip:** You might critically evaluate the ETF as a form of *ineffective* diversification, in the sense that the basket of securities by definition faces all of the same types of risk exposure.

Some energy subsectors could suffer on an ETF level as well. For example, alternative energy stocks could decline within an ETF framework if clean, efficient, and abundant sources of fuel were discovered and developed. However, analysis of energy-based ETFs shows that many of these funds are diversified among the energy subsectors, so they contain the desirable built-in diversification that adds protection to a portfolio.

No one should invest in an ETF, commodity index fund, stock, or in options, without knowing the components of risk and ensuring that those attributes match well with levels of risk tolerance. An unfortunate reality is that many investors do not perform adequate analysis but tend to make decisions quickly and often without knowing the effects of economic, political, or even market forces on their decisions. For example, investor confidence rises with the averages, even though those averages do not represent any one company or its stock. The Dow Jones Industrial Average (DJIA) is thought by many the represent the entire market even though it contains only 30 stocks. Only two energy companies—Chevron and Exxon Mobil—are included in the index. Together, these represent 10.73% of total weight as of May, 2013.[5]

Many investors judge the market's health based on movement of the index, but this can be misleading. For example, does an upward direction in the DJIA justify deciding to buy energy stocks? No. With only 10.93% of the total DJIA weight, the two companies in the index represent a small segment of the energy market and an even smaller segment of the stock market—two out of thirty stocks among thousands that are listed and only 10.73% of the index itself. Even so, people decide to move in or out of the market based on what happens with the DJIA.

It is wise to consider and analyze the fundamentals of the specific sector before making a decision to buy or to sell and to then study individual stocks in terms of financial strength, profitability, long-term trends, and competitive stance. This all may occur regardless of what is taking place on the index side of the market; however, index trends do tend to affect all listed companies. Regardless of fundamental strength or weakness, the tendency of a majority of stocks to move with the market should not be ignored. It makes no sense to buy or sell in isolation from market-wide trends, but it also makes no sense to invest in one company based solely on what the index movement tells you.

CHAPTER 10

USING A BROKER VERSUS GOING ON YOUR OWN

> We're drowning in information and starving for knowledge.
>
> Rutherford D. Rogers, in *New York Times*, February 25, 1985

AS AN ENERGY INVESTOR, DO YOU NEED A BROKER, FINANCIAL planner, or financial professional with any other title? Some people do and some do not.

A related question: Should a financial professional be an expert in the energy market? It would probably be difficult to find an advisor who is such an expert, and it would not hurt for your advisor to be well versed in the special fundamental and technical aspects of investing in the energy market. But is it necessary? No. If you were investing in gaming industry stocks, would you seek a financial advisor who visits Las Vegas and plays blackjack every month?

It's a misconception to assume that a financial professional has to be an expert in all aspects of what you need and want. It's much more important to find someone who understands market risk and who knows how to match products to acceptable risk levels. It's even more important to find a professional who will listen to you and respect you rather than try to talk you into something you don't want.

Key point: It is far more important for a financial professional to understand how to manage risk than to know all about a particular market (such as energy).

An analogy makes this point. Smart car salespersons know not to argue with the customer. If you go to the car lot and tell them, "I want a van because I have a family and we need to have room for the kids," that's a pretty clear communication. You expect to be shown some choices in vans. But one response might be, "No, you don't want a van, you want this subcompact because it's fast and snazzy." This tells you that the salesperson is just not listening. The same rule applies to financial professionals. Clients know that they're dealing with someone who is selling products or services, but there are many ways to sell. When it comes to something as personal as your financial health, you expect and deserve to define the field of play and to express your goals, risk tolerance, and preferences. If you're conservative, you don't want to be talked into high-risk speculative products. It's the van and subcompact all over again, only in a different environment.

You don't expect your financial professional to know everything about the energy market or any other sector in which you invest. But you do have a set of expectations. In deciding which financial professional to retain, it is important that the following attributes are present:

1. *Access to specialists.* No one can be expected to know about everything affecting you and your investing decisions. To know all about taxes, estate planning, insurance, legal questions, stocks and bonds, options and futures, and so much more that affects how and why you invest would require one person of amazing diversity—one that does not exist. An effective financial professional is able to directly address many of your needs but also has access to associates with narrow specialization. This access may be through other experts in the same office or to experts on the outside. Anyone who claims to be able to provide you with everything you need is not a good prospect and cannot possibly serve you well.
2. *Knowledge about investing goals, not necessarily in specific products.* The true professional is one who knows how to match your goals and risk tolerance limits with appropriate investments. This does not mean the professional needs to know the details of products; it does mean that you can rely on that person's skill in translating your goals and risk tolerance into a

range of choices. So much emphasis is placed on products that individual goals are lost in the process. Any time a financial professional makes a specific recommendation without knowing all of the details of your personal goals and investing views, be very cautious.

3. *Competent listening skills.* Does your advisor tell you what you want, make fast recommendations, or rush you to accept suggestions? Or does that individual listen to you, ask for more information, and respect you enough to learn about your goals, biases, and wishes? Listening skills distinguish an effective financial professional from those who are merely salespeople trying to close a sale.
4. *Experience.* Everyone has to start somewhere, and even financial professionals need to start out as apprentices to more experienced leaders and mentors. If you are going to trust someone to work with you to find appropriate investments, that person must have the experience and knowledge to do so. If you are assigned to a new financial professional who just got licensed but has no job experience, how can that person know what you need? Insist on working with an experienced professional. Since you are going to pay for the service, you have the right to demand that your financial professional has the experience to give you sound ideas.

Tip: Finding the right financial professional is the same process as selecting any other type of professional. Compare, ask questions, and find the best *individual* match.

THE ROLE OF A BROKER

Distinctions have to be made between financial professionals with various titles.

A "broker" is usually an individual working in a brokerage firm. Brokers execute orders placed by customers in exchange for a commission or fee. By this definition, you do not expect a broker to have any specific expertise but only to execute your orders.

Some traders interested in stocks will pay brokers for advice. However, any time you take advice from someone, you should first understand that person's source of information. Has a broker performed extensive research about a company? Or is the advice based on what the firm underwrites or promises to move to the market?

A brokerage firm is involved not only in order placement for customers, but also in underwriting, the movement of shares of new issues to the market. Thus, when a broker contacts a customer and recommends a specific stock, what is the motive? If that firm acts as underwriter for the new issue, then every broker in the firm is expected to market shares to clients. That is, the company is not necessarily appropriate, safe, or potentially profitable, and the reasons your broker contacts you relate more to the underwriting activities than to any help you expect to get for the fees or commissions you pay.

The brokerage firm accepts a number of shares and also assumes the risk of marketing those shares for the issuing company. The brokerage underwriting this sale is compensated, and this often is the largest source of revenue for the brokerage firm. Brokers are also motivated to sell shares to customers as quickly as possible; the longer it takes to sell shares, the greater the risk that share price will decline, thus eroding the brokerage's profits. This activity, given the dignified name, investment banking, is often just a hard sell of securities to the brokerage's customer base. One problem with using a broker then is that when recommendations are made, you do not know whether your broker is working on your behalf and trying to find suitable stocks for you or is working for management in trying to market new issues.

Tip: Recognize a conflict of interest whenever you are given advice. Ask the right questions, especially when that advice involves placing your money at risk.

As an investor or trader in the energy market, you might discover that a broker tries to convince you to buy shares of new issues, whether in the energy market or not. By asking the following series of questions, you can determine whether your broker is working for you and listening to what you want:

1. Is this a new issue?
2. If so, is your firm an underwriter of the stock?
3. What fundamentals have convinced you that this is a good price?
4. Why is this stock a positive match for me (based on fundamentals and risk)?

You might discover that a broker is not prepared to answer these questions. In that case, you may consider looking elsewhere for advice or even performing your own research and using a brokerage firm solely for trading. In that case, look for a discount broker to avoid paying higher commissions for advice you don't need—or worse, for advice not given in *your* best interest.

A great problem with brokers is how well the public understands these different roles. The Securities and Exchange Commission (SEC) performed an analysis of this problem and summarized the key issue as follows:

> Retail investors seek guidance from broker-dealers and investment advisers to manage their investments and to meet their own and their families' financial goals. These investors rely on broker-dealers and investment advisers for investment advice and expect that advice to be given in the investors' best interest. The regulatory regime that governs the provision of investment advice to retail investors is essential to assuring the integrity of that advice and to matching legal obligations with the expectations and needs of investors.[1]

The study conducted by the SEC concluded as well that:

> Many retail investors and investor advocates submitted comments stating that retail investors do not understand the differences between investment advisers and broker-dealers or the standards of care applicable to broker-dealers and investment advisers. Many find the standards of care confusing and are uncertain about the meaning of the various titles and designations used by investment advisers and broker-dealers. Many expect that both investment advisers and broker-dealers are obligated to act in the investors' best interests. The commission has sponsored studies of investor understanding of the roles,

> duties, and obligations of investment advisers and broker-dealers that similarly reflect confusion by retail investors regarding the roles, titles, and legal obligations of investment advisers and broker-dealers, although the studies found that investors generally were satisfied with their financial professionals. Several of the recommendations listed below are designed to address investor confusion and provide for a stronger and more consistent regulatory regime for broker-dealers and investment advisers providing personalized investment advice about securities to retail investors.[2]

Most retail investors are aware only of the broker's role in making investment recommendations and then placing orders. The underwriting role is not as widely known or understood, and brokers do not necessarily disclose to customers that their firm acts as underwriter on issues they are recommending for purchase. The mixing of their role as securities advisory service and as underwriters is commonly not disclosed to the customer, and this is a problem for customers relying on a broker's advice. Confusing this issue further, a broker might advise a customer that the firm is acting as "market maker" for a new issue. This means that the brokerage has bought a block of shares and holds these shares in inventory for the purpose of selling those shares to its customers. This distinction relates only to how shares are held; the brokerage is still acting as underwriter, whether it is acting as market maker or not.

Key point: A majority of traders relying on a broker for investment advice are simply unaware of the broker's potential underwriting role and related conflict of interest. Knowing this makes you a more informed, smarter investor.

THE FINANCIAL PLANNER OR PROFESSIONAL

Beyond the role of broker—restricted to recommendations and execution of trades—is the financial professional. This individual is more likely to provide a range of investment and financial services, to hold advanced licenses, and to work for an advisory fee rather than for a commission. The broker, in comparison, is normally compensated by commission based on the number of shares and dollar value traded.

Anyone working with a financial advisor should expect personalized service based on (a) well-understood risk tolerance, (b) personal investing goals, and (c) personal preferences including biases toward specific markets.

For example, a financial professional should be expected to provide a broad range of services tailored specifically to each client. This invariably brings in a multitude of specialized disciplines, either directly or through associates. Financial professionals are likely to define the range of their services in the context of a "value proposition." This is a concise series of statements about who the financial professional is, what services are provided to clients, and how those services are delivered. In the training and orientation of associates, financial firms devote classes and exercises to helping financial professionals articulate their value proposition, and the purpose is to focus on self-definition on behalf of the client. A properly articulated value proposition helps financial professionals to:

- set themselves apart from their competitors in the personalized services offered,
- gain market share in the demographics they target,
- improve the quality of services offered, and
- augment efficiency in arriving at and defining customer needs.

As you search for a financial professional, remember that expertise in the energy sector does not have to be a prerequisite. It makes more sense in terms of "target marketing" for a financial professional to know the attributes of customers and what they seek. For example, financial professionals who want to work with individuals interested in investing for current income should be aware of the many ways this goal can be accomplished. Those focusing on value should be aware of what constitutes "value" in the market. As an energy investor, you might want to work with a financial professional who can help you develop a portfolio that diversifies and reduces your risk exposure.

The value proposition is intended as a communication tool, a method for professionals to tell their clients who they are and what they offer. It can be a very simple statement or a more developed series of definitions. The following is an example of a simple value proposition:

My clients are self-employed business owners intent on preserving capital through focus on conservative but high-yielding investments in specific market sectors: agriculture, energy, and housing.

This describes an investor who understands the three basic necessities (food, energy, and shelter) and who wants to focus on those sectors. However, the customer base is more narrowly defined as business owners. This implies that this financial professional will also be interested in assisting clients with employee benefits, pensions or profit-sharing plans, insurance, and other related financial services that go beyond investing.

> **Tip:** The value proposition is a set of carefully thought-out definitions of the individual, what is provided, and to whom.

As a simple statement of (a) who the customers are, (b) their primary investing goals, and (c) what the financial professional does to help, the above value proposition is simple and direct. As opposed to offering a range of services (which is vague and unfocused), this value proposition is very specific regarding clients, goals, and the potential range of services.

The value proposition (which may also be defined as a "promise of value the financial professional delivers to the customer") can be applied to the individual, an entire firm of professionals, or to a specified range of products or services. It is found not only among financial professionals, but in a wider range of providers, usually those involved in marketing or sales. Another way of expressing this is: "Strategy is based on a differentiated customer value proposition. Satisfying customers is the source of sustainable value creation."[3]

The value proposition may also be a much more expansive set of statements that summarize the benefits a firm or individual offers. A "Value Proposition Builder" is one model containing six stages: market, customer experience, offerings, benefits, alternatives, and proof.[4]

Another approach is to develop a "core value proposition," a multistage series of definitions concerning (1) services and customers to whom they are provided, (2) definitions of wealth, (3) the process, (4) what the individual professional brings to the picture, and (5) other unique attributes customers need and want.

Services and customers are included in a defining statement. For example, a financial professional might define a range of services as "building and protecting wealth through appropriate products and planning," and clients may be defined as "a narrow group of high net worth individuals, small business owners, executives, and professionals." The complete "service/customer" statement then is: "I help a select group of high net worth individuals, small business owners, executives, and professionals to build and protect wealth through appropriate products and planning."

Definitions of wealth should include more than products. If your financial professional offers only a narrow range of investments, then you are not being fully served. The core value proposition should identify all areas of wealth building and protection, which fall into broader ranges: investments, insurance, banking/credit, and estate planning. A truly comprehensive financial plan will address these four areas. As an energy investor, you want your investments to fit into the first classification, but you must also consider measures needed in the other three areas of the definition of wealth. The value proposition statement may now be expanded to read: "I help a select group of high net worth individuals, small business owners, executives, and professionals to build and protect wealth through appropriate products and planning. This includes risk management in four areas: investments, banking/credit, insurance, and estate planning."

The process refers to defining how the financial professional helps you. It's not enough to push products and to get you to sign on the bottom line; the services should be designed for you. Therefore, a statement is added to the value proposition. For example: "I engage in a process based on listening to my clients to discover their experiences, knowledge, background, and goals."

Individual professional attributes are what professionals offer to you to make their services especially appropriate. For example, the value proposition continues: "I offer years of experience, training, and knowledge, so that my skills and the skills of my associates are put to work for every client and are focused on attaining their goals."

Other unique attributes include any special aspects of the professional's focus and the services offered. This might include a background in energy investment (for example, the financial professional might have been an executive in an oil company or even worked in an oil field) or a specialization in women's financial plans or specialized financial planning for sports professionals, doctors, lawyers, or any other profession. You may find a general practitioner or someone who focuses narrowly on one segment of the market. For example, if you are referred to a financial planner with a background in energy, that person's value proposition might conclude with: "My experience as marketing vice president of an energy company qualifies me to address investment and other needs of those drawn to this market sector."

> **Tip:** The end result—the value proposition expression—may look simple and perhaps it should. But it is the result of a lot of thought and energy.

Now the complete value proposition is done and reads as follows:

> I help a select group of high net worth individuals, small business owners, executives, and professionals to build and protect wealth through appropriate products and planning. This includes risk management in four areas: investments, banking/credit, insurance, and estate planning. I engage in a process based on listening to my clients to discover their experiences, knowledge, background, and goals. I offer years of experience, training, and knowledge, so that my skills and the skills of my associates are put to work for every client and focused on attaining their goals. My experience as marketing vice president of an energy company qualifies me to address investment and other needs of those drawn to this market sector.

When you see this kind of short but descriptive statement from a financial planner, you have a clear and fully defined explanation of how the individual works with clients. The final step in how to find a financial planner based on personality is the question of whether the individual *sells* or *listens*. Those who let you explain your own perspectives on your financial life and goals are more likely to offer you products tailored to you and not a predetermined set of products they want to sell. Consider two versions of a first meeting's dialogue:

Sales focus:

Financial professional: So, let me tell you about a very interesting new mutual fund that has an amazing record of profits in its first year.

Client: Actually, I wanted to talk to you about energy ETFs and commodity index funds. This is the way I—

Financial professional: You don't want to do that; it's too narrow. A mutual fund spreads your risk over a broader market. Forget about ETFs; you can do better.

Listening focus:

Financial professional: So, tell me, what kinds of investments appeal to you? I'd like to start there and then expand into a discussion of other financial needs.

Client: I think investing in the energy market would be smart. I've done some reading on ETFs and commodity index funds.

Financial professional: Tell me more...

The difference in these two dialogues is glaring. Many who have visited sales-oriented financial professionals have experienced the first version many times, but very few people have had exposure to someone who truly listens to them and who encourages sharing of more information. The second version is better communication; it is also a better way to build dialogue and trust between two people.

GUIDELINES FOR USING BROKERS AND FINANCIAL PLANNERS

Professional advisors may explain investment choices to clients, help them develop a sensible financial plan, identify different risk levels, and pick from among many different choices. In the energy sector more than most other sectors, these choices can be very complex.

There is an advantage to paying for professional advice. Even if a particular financial professional is not knowledgeable about investments in the energy sector, you would expect expertise in various types of products, knowledge about market risks, and the ability to arrive at a series of recommendations based on your risk tolerance.

> **Tip:** When you pay for professional advice, be sure you know what you're paying for. You are not going to get access to a secret formula for wealth, but you should expect good advice. A financial professional adds value by helping you navigate through your choices.

The Financial Industry Regulatory Authority (FINRA) has established a set of rules about conduct of financial professionals, mostly commonsense rules for ethical behavior. This serves not only as a means for guiding conduct, but it also establishes a basis for complaints and remedial action when these rules are violated.[5]

With this in mind, remember that titles such as financial advisor, analyst, consultant, planner, and manager are generic and are not credentials following any training; they are not accredited designations or licenses. In this business titles have to be investigated thoroughly. The following is a brief list of guidelines for selecting candidates to work with you:

1. *Know your goals and preferences in advance.* A common error people make is to visit a financial professional without having any idea of what they want. Most people know that if you visit a car lot with no budget and no preference, you are likely to be sold something you don't really want. The same rule applies to all examples of working with others. When you are ready to interview a financial professional, first decide what you want to accomplish. What are your goals? Are you trying to build your net worth, save for retirement, start your own business, or preserve your capital? How conservative are you? Do you seek investments in a particular market sector (such as energy)? If so, have you investigated the methods available to you (futures, stocks, ETFs, commodity index funds)? The more prepared

you are, the more successful you will be in working with a financial professional.

2. *Ask for referrals, especially from other professionals.* Remember, a referral from someone who doesn't know much about finance is not going to be very useful. Someone may tell you his or her financial adviser is "the best in the business," but what does that mean? It probably is not worth much; however, a referral from your accountant, attorney, or other professional may be based on a higher standard, so you should look to these professionals for referrals.
3. *Meet with and interview several candidates.* Don't settle for only one referral. Get several and then meet with each of them. Inquire about education, professional designations, licenses, and experience.
4. *Check out each person you interview.* Make sure each candidate's licenses and designations are the real thing and currently in effect. Also check whether any complaints have been filed with associations or regulatory agencies and whether any disciplinary actions were taken.
5. *Ask how the financial professional is compensated.* An individual who is paid by commissions is paid by the mutual fund or other sponsor where your money is invested, and of course that payment comes out of your account. Financial professionals paid by commissions have a conflict of interest because they will invariably refer you to products that pay a commission (such as a load mutual fund versus a no-load fund, for example). The advice you get is likely not objective.

   Financial professionals operating on a fee basis charge a fee, usually based on an hourly rate or a flat fee. This might seem expensive, but with a commission-based adviser you are still paying, and a fee-based adviser will likely be more objective.

   A troubling situation is one in which a financial professional gets both a fee and a commission. You pay twice in this case—first, you pay the fee for advice and then the commission is deducted from everything you invest. Recognizing that it is a conflict to collect both, some financial advisors have set up a separate entity to collect commissions. For example, as an individual they charge a fee, and then their corporation (which they own 100%) gets any commissions earned from the advice you are provided. But you are still paying twice, so if the person you interview explains that he or she has things set up this way, remember that it's still a double payment.

5. *Make sure the financial professional is properly registered.* Several registration requirements are imposed on financial professionals. At the very least, make sure they are registered with FINRA, the SEC, and your state's securities regulator. The firm should also be a member in good standing with the Securities Investor Protection Corporation (SIPC). Professional who will also sell insurance must be registered with the state insurance commission. They may also have to file as registered investment advisors (RIA).
6. *Check licensing levels through FINRA.* There are three primary licenses most financial professionals hold based on the type of business they conduct. These are:
   - Series 6, limited investment securities. This license allows the financial professional to sell mutual funds, variable annuities, unit investment trusts, and other "conduit" (pass-through) products.
   - Series 7, general securities representative. This license allows the professional to sell individual securities including stocks, options, bonds, and mutual funds. Excluded are commodities, real estate, and insurance. Those who pass the Series 7 exams are usually referred to as "registered representatives."
   - Series 3, covering commodities. Anyone with this license can sell futures contracts directly and may also be involved in marketing options on futures.

   A long list of additional FINRA licenses is also available for more specialized transactions. Make sure you know which licenses a financial professional holds. But don't mistake licenses for experience and expertise. The license is only a starting point.
7. *Consider limiting your search to those who are Certified Financial Planners (CFPs).* Becoming a certified financial planner requires extensive education and examination. These professionals also must agree to abide by the Rules of Conduct and Code of Ethics imposed by the CFP Board. To get the designation, applicants have to go through what are called the four Es:[6]

a. *Education*—the CFP Board requires completion of college-level courses as well as a BA or higher degree.
b. *Exam*—the certificate examination ensures that applicants have mastered all the topics required to serve as a financial professional.
c. *Experience*—every CFP must spend three years as a financial planner, including at least two years as an apprentice.
d. *Ethics*—the Standards of Professional Conduct is based on a background check for criminal, civil, and other problems, and this is extensive.[7]

The CFP Board also documents any complaints against its members, so that checking on someone with whom you are thinking of working is easy; the website is a good starting point to verify that the person is a CFP and whether or not any complaints or disciplinary actions are on file.

Key point: The best way to find a competent professional is to thoroughly check out any candidates. Too many investors are overly trusting and as a result might end up with someone who does not know about risk management.

The greatest disadvantage in relying on advice is not cost (although cost is one factor). A more serious potential problem is in determining whether or not you are getting objective advice. A starting point is to ask yourself whether you need the advice of an expert (assuming you are able to find an expert who is also objective). Many investors decide to make their own decisions, perform their own research and analysis, and rely on expertise of others only when specific situations arise (legal questions, research about a specific company, or technical issues in the energy industry, for example).

OUTSIDE HELP AND THE BIGGER PICTURE OF LONG-TERM PLANNING

Personal financial planning should be highly individual. Any planner who provides clients with a "cookie cutter" version of a financial plan is not providing real service. Your long-term plan should consider all aspects of your financial life: a stock or other market portfolio, insurance, long-term planning, risk identification, estate planning, tax planning, and more. Only when this is all in place can you know how to build the best portfolio and decide whether investment in the energy sector should be part of it.

As with any type of personal service, you should expect to get *personalized* attention geared toward you and your risk tolerance, your goals, and your long-term desires. The ideal candidate will be a financial professional who understands that is as a priority. Your financial professional does not need to have an in-depth understanding of the energy sector; however, knowing how to analyze both fundamental and technical attributes of that market is essential. This applies to all sectors, however. If your portfolio is diversified among five different sectors, you cannot expect your adviser to be an expert on all five—or even on any one of them. You do need someone who knows how to analyze risks and match them to your risk tolerance, who can identify inherent risk levels in specific product lines, and most of all, who knows that listening is more important than talking.

Key point: Listening skills are essential in a professional. Fortunately for you, when someone lacks these skills, you can identify that problem very quickly.

A final thought about financial professionals: Just as you bring your bias to the table when you invest, so does everyone else. You might be close-minded about futures and options, and more likely to prefer stock ownership, ETFs, or commodity index funds. A financial professional who listens will understand these biases and respect them and work with you to find products that are a good match for you.

Problems occur when the financial professional's bias is different from yours and too much time is spent trying to "sell" you on an idea you simply don't like. For example, if the financial professional says, "You should take another look at options. You can make a lot of money trading them," that tells you they are not listening. A more enlightened message would be, "I understand that you don't want to trade options, so let's work to find products you're comfortable with and that will meet your goals."

That's the person you might just want to work with because the message demonstrates keen *listening* skills.

CHAPTER 11

FUNDAMENTAL ANALYSIS OF THE ENERGY MARKET

Advice is not disliked because it is advice, but because so few people know how to give it.

Leigh Hunt, *The Indicator*, 1821

THE SYSTEM OF FUNDAMENTAL ANALYSIS EXISTS ON TWO LEVELS FOR the energy market, and both levels are not always well understood.

The first level is the one known best: the financial history and trends as reported by companies to their stockholders. Balance sheets and income statements are the main vehicles of communication. The balance sheet reports balances of assets, liabilities, and net worth as of an ending date (end of the fiscal year, for example), and the income statement summarizes transactions over a period of time (the full fiscal year, for example). When these are published, the balance sheet ends on the last day of the period while the income statement summarizes the revenues, costs, expenses, and profit over the entire period.

The second level of analysis might not even by acknowledged by many investors as fundamental because the components are so widely discussed in the market and for stocks, sectors, and the economy as a whole. This level includes all of the economic, geopolitical, and other factors affecting the energy market. For most sectors, the economic and geopolitical fundamentals are fairly straightforward. Supply and demand are based on how well a company competes in the sector, and this is influenced by currency exchange, trade limitations, and other market factors.

Key point: In the energy market, *perception* is at least as important as reality. You see this time and again in how the fundamentals affect price.

In the energy sector, however, a distinction has to be made: The *perception* of the market often is more influential than the facts. This is true in several ways:

1. *Energy products have a negative impact on the environment.* The belief that fossil fuels contain pollutants has been held for decades. However, with improved emission standards over the past half century, the threat today is lower than in the past. Even so, the perception that fossil fuels do damage to the environment—specifically to air quality—is widely accepted, and it affects the prices of fuel. This perception is a key fundamental with several outcomes, including the rise of clean energy alternatives. This occurred with equal fervor in the United States during the 1973 Arab Oil Embargo.[1]
2. *Energy products accelerate or even cause climate change.* The belief in negative environmental impact pales next to the accusation that energy emissions cause or worsen global climate change. Whether this is true or not does not matter; perception is the more important factor here. Although the science is controversial, the idea that fossil fuel is creating climate change also supports the search for clean alternatives; however, as every industry and every household continues to rely on energy, the likelihood of replacing fossil fuels entirely is slim. The most likely outcome will be higher prices for fuel.
3. *The world is running out of fossil fuels.* This idea has been around for over 100 years, ever since the beginning of the auto age and the age of air travel. In spite of repeated estimates of world fuel running out within a decade or less, new reserves are being found constantly, and new and more economical methods of extraction have made reserves available that were once unfeasible. The United States has decades of natural gas reserves, but conversion of dependence from oil to natural gas is not a simple matter. The repeated estimates of oil's production peak have all turned out wrong. Today,

instead of estimating the future peak year, many claims are being put forth that oil has already peaked and that production is on a decline. One article cited a belief that "there is no longer any spare capacity to respond to increases in demand" while also pointing out that some production is on the rise. For example, BP "oil production was actually more than 82 million barrels per day in 2010, higher than the proposed plateau of 75 million."[2]

In fact, another source observes that "New oil sources, many of them unlocked by new technology—the Canadian oil sands, tight oil in North Dakota and Texas, ultra-deepwater oil in the Atlantic—has [*sic*]) helped keep the supply of oil growing, even as greater efficiency measures and other social shifts have helped blunt demand in rich countries like the United States."[3]

4. *The United States gets most of its oil from the Middle East.* In fact, as shown in chapter 1 and figure 1.5, the United States produces 39% of its oil domestically and gets another 35% from Mexico and Canada—a total of 64% from North America. Only 13% comes from Persian Gulf imports, and another 13% from elsewhere. This perception of our dependence on the Middle East is distorted but widely believed. Like so many false beliefs, it is also persistent. The fact is that if the United States were to exploit all of its known reserves of oil (including shale and sand oil), natural gas, and other forms of fossil fuels, it would be possible and realistic to become self-sufficient. The belief that the United States gets most of its oil from Persian Gulf members of OPEC has a clear and significant effect on oil prices, in spite of the true economics of this sector.

Defining Fundamental Analysis

The fundamentals in their direct form (related to individual companies) include financial information, used to identify investment candidates and to decide whether to buy or sell shares of stock, ETFs, indexes, or futures contracts. The range includes quarterly financial reports, annual statements, reports filed with the SEC, and the study of trends and indicators.

Key point: Fundamentals are potentially complex, but the purpose of analysis is to boil down the numbers into recognizable trends.

In this form, the fundamentals are used primarily as a means for picking stocks. In chapter 6, three energy companies (Exxon Mobil, Chevron, and ConocoPhillips) were evaluated using a short list of fundamental indicators (PE ratio, revenue and profits, dividend yield, and debt ratio). This section provides a broader view of how *trends* are used in fundamental analysis and applied to the income statement.

Trends are more often associated with technical analysis, where price and volume are charted and analyzed to spot changes in momentum, evolving moving averages, and reversal signals with confirmation. But trends are also used in the fundamentals. The previously discussed P/E ratio, revenue and profits, dividend yield, and debt ratio are among the most important trends to follow; in addition, you gain fundamental value by tracking tangible book value per share, earnings per share, and common equity.

Tangible book value per share is the calculated value excluding intangible assets, such as value assigned to goodwill, intellectual properties (patents, trademarks, copyrights), and similar kinds of assets. To calculate this, first deduct intangible assets from total assets and then divide the result by the number of common shares outstanding:

(total assets – intangible assets) ÷ outstanding common shares

The resulting value per share is a key fundamental indicator, but like most such indicators a current intangible book value per share (IBVPS) is of little value. The *trend* in this calculation is more revealing, especially when comparing the valuation for two or more companies. It is a type of worst case calculation, the value per share in the event of total liquidation. In that instance, all of the tangible assets would be disposed of, and although a company may be quite solvent, the calculation provides a means for monitoring changes in asset valuation over many years. It addresses the effectiveness of the company's ability to build value over time.

For example, Exxon Mobil (XOM) reported a strong and consistent record of growth in IBVPS over a ten-year period, starting at \$13.69 in 2003 and ending at \$36.84 by the end of 2012, a growth of 269% over the decade. This ten-year trend is summarized in figure 11.1.

Note that XOM's record moves consistently higher over the decade. This reflects the ability of management to "grow" assets even through rough periods (such as 2009–2011). Although this is a subtle indicator in terms of comparative fundamental strength or weakness, this level of consistency serves as an excellent test for a comparison between two companies in the same sector.

> **Tip:** When you cannot decide between two excellent stocks, favor the one with the strongest and most consistent long-term trends.

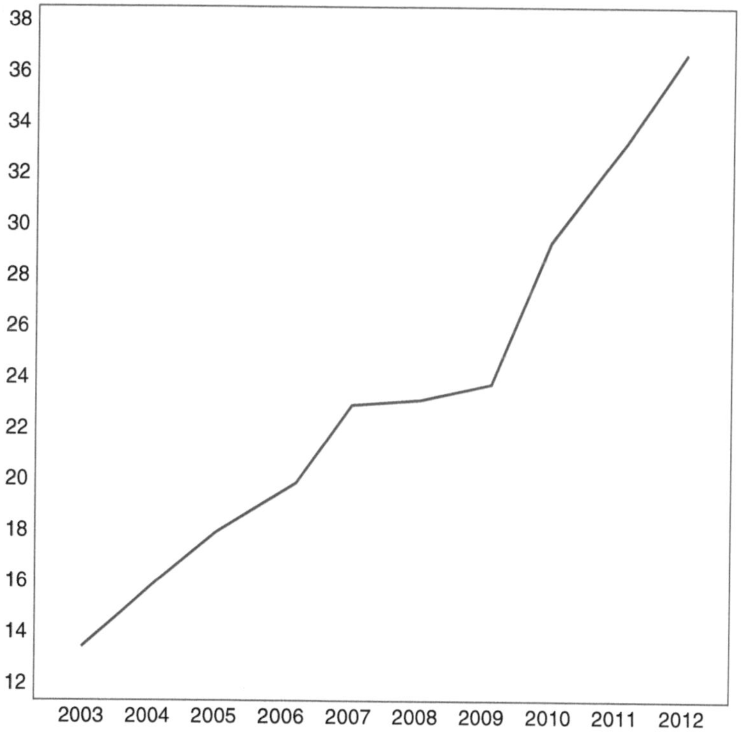

Figure 11.1 Exxon Mobil, tangible book value per share, 2003–2012
Source: Prepared by author from raw data at S&P Stock Reports

For example, the same chart for Chevron reveals an even more consistent upward curve in tangible book value per share. Like XOM, the CVX record continued upward even through the difficult market of recent years. This is summarized in figure 11.2.

The path is not always this smooth for energy companies. For example, ConocoPhillips reflected what was going on in the market more than its two competitors. The decline at the beginning of the market adjustment was noticeable; however, the 2012 decline, which was quite severe, was not as easily explained. Nevertheless, as a point of comparison, tangible book value per share provides excellent fundamental comparative information. This is shown for COP in figure 11.3.

A better-known and more widely followed fundamental indicator is earnings per share (EPS). You cannot accurately compare earnings per share on a dollar value basis between companies, because the value of EPS is unique to each company. For example, one company might earn $45,800,000 in a year, while another earns $22,900,000. However, based on how many shares were outstanding, no valid

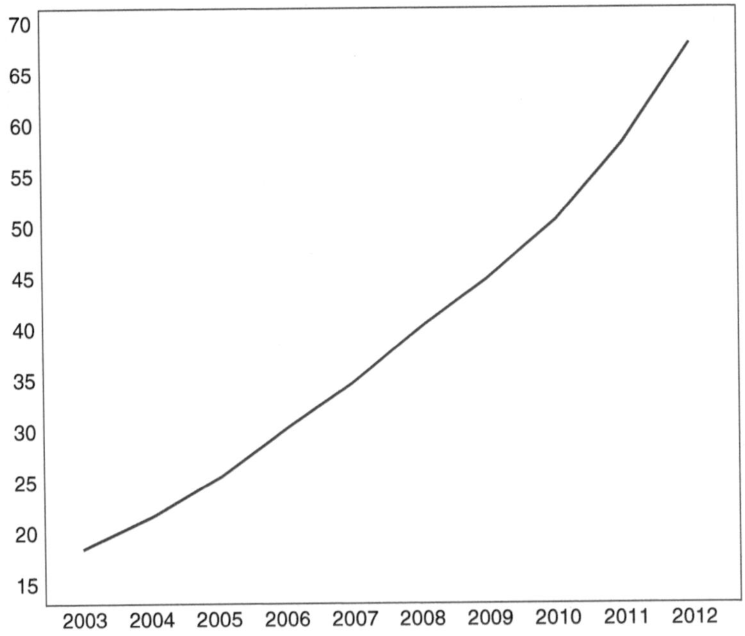

Figure 11.2 Chevron, tangible book value per share, 2003–2012
Source: Prepared by author from raw data at S&P Stock Reports

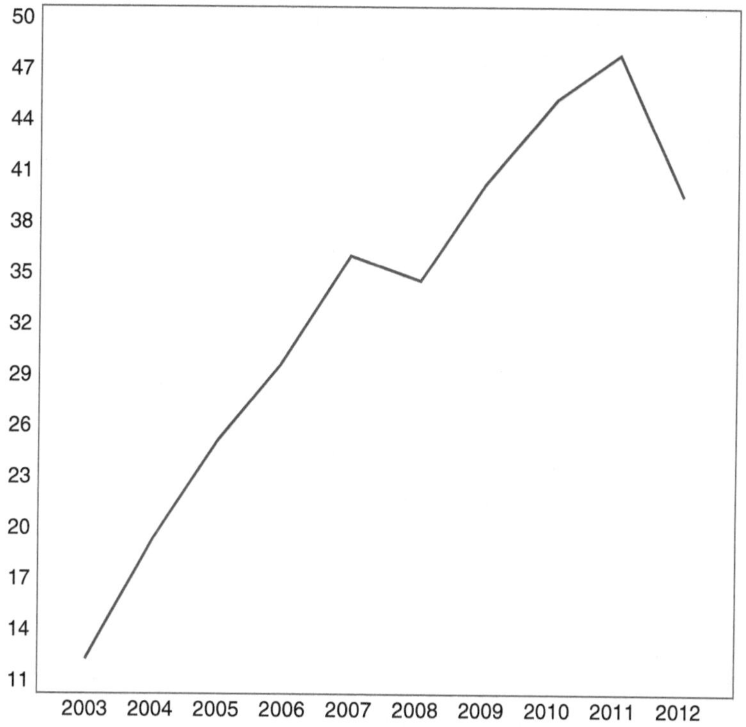

Figure 11.3 ConocoPhillips, tangible book value per share, 2003–2012
Source: Prepared by author from raw data at S&P Stock Reports

comparison can be made. If we assume that the first company had 20 million shares outstanding and the other had 10 million, both had EPS totals of the same dollar value:

$45,800,000 ÷ 20,000,000 = $2.29
$22,900,000 ÷ 10,000,000 = $2.29

Although the first company earned twice as much, its EPS was identical to that of the second company with half the earnings but twice the number of shares.

To calculate EPS, divide net income by the *average* number of outstanding shares. This is not as simple as it might sound, because the number of shares may change often during a fiscal year. This is due to new issues of common stock or to share retirement through buyback by the company. So the average is not just the beginning balance added to the ending balance, divided by two. For the accurate average you have to consider the number of months involved as well. Some versions of the calculation just use the ending number of shares in place of the average. This could be seriously inaccurate, however, if the number of outstanding shares has changed substantially during the year.

The most important point about this is to make sure that the calculation of EPS is consistent over many years for a company, so that the trend is reliable. The basic formula is as follows:

total net earnings ÷ average outstanding common shares

EPS is further changed when convertible bonds or preferred stock is taken into account. This *diluted* EPS is more accurate, but it may create inconsistencies if previous years' EPS did not include convertible instruments.

Key point: The calculation of EPS may be based on several variables. The most important rule is to calculate it consistently so that your picture of trends is realistic and accurate.

EPS reported to the SEC (also called GAAP EPS, with GAAP referring to Generally Accepted Accounting Principles) included nonoperating revenues such as capital gain from the sale of assets, currency exchange profits, or interest income. In comparison, ongoing EPS includes only core earnings from operations and excludes one-time and nonoperational income. A third version is called *pro forma* EPS and is based on estimates of future net income. This is an uncertain version of EPS because the basis is an estimate and not actual numbers. A company may also calculate "cash EPS," which uses operating cash flow instead of net earnings. In this version, noncash expenses such as depreciation and amortization are excluded.

Accountants and analysts argue over these distinctions. However, if you are an investor attempting to track a company's EPS, the most important thing for you to remember is that over a period of years, the EPS number used should be calculated consistently. Otherwise, the trend will be distorted and not reliable.

This is a trend that, like so many, is best tracked individually for a company. You cannot compare EPS between two or more companies because their numbers of shares make such comparisons unreliable. For example, the number of shares outstanding varies greatly among the three energy companies discussed in this chapter (Table 11.1).

Exxon Mobil reported earnings per share rising strongly for six of the past ten years and then a significant decline in 2009, typical of the overall market at that time. The EPS record is summarized in figure 11.4.

Table 11.1 Market cap and outstanding shares

Company	Market cap	Outstanding Shares
Exxon Mobil	$408.53 billion	4.45 billion
Chevron	242.35 billion	1.94 billion
ConocoPhillips	76.45 billion	1.22 billion

Source: New York Stock Exchange

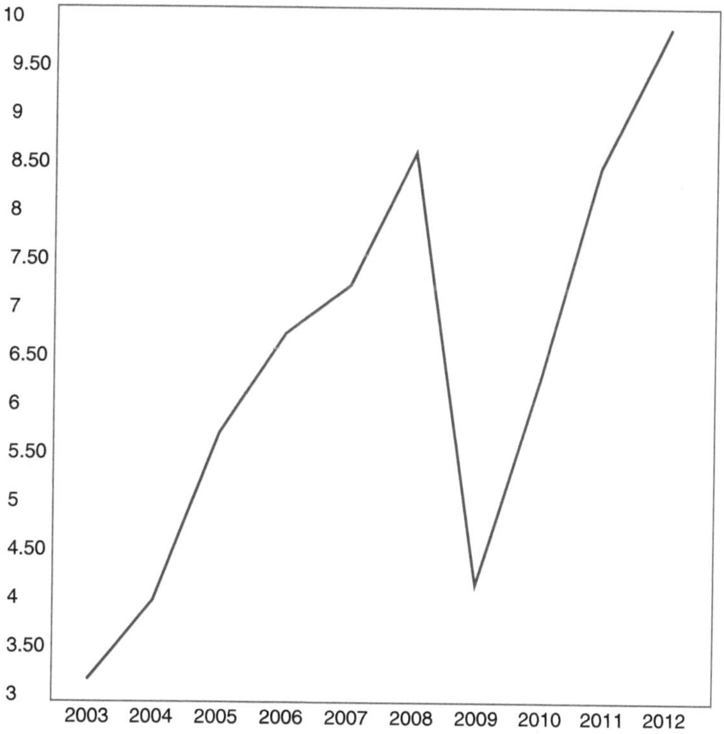

Figure 11.4 Exxon Mobil, earnings per share, 2003–2012
Source: Prepared by author from raw data at S&P Stock Reports

Chevron's EPS was also reflective of market conditions during the decade. You might take comfort in seeing that both XOM and CVX acted and reacted in a similar way. This reveals that both energy stocks responded to market conditions (both the overall market and the energy sector) in very similar ways. The CVX ten-year record is shown in figure 11.5.

The third company, ConocoPhillips, differed greatly from its two competitors. Its volatility was more extreme, with earnings dipping to a loss of -11.16 per share per share in 2008 and dipping once again in 2012. COP is far more volatile than both XOM and CVX. The record for COP's ten-year EPS is shown in figure 11.6.

> **Tip:** Extreme volatility in fundamental indicators tells you that predicting future changes is going to be difficult. You cannot depend on the trend.

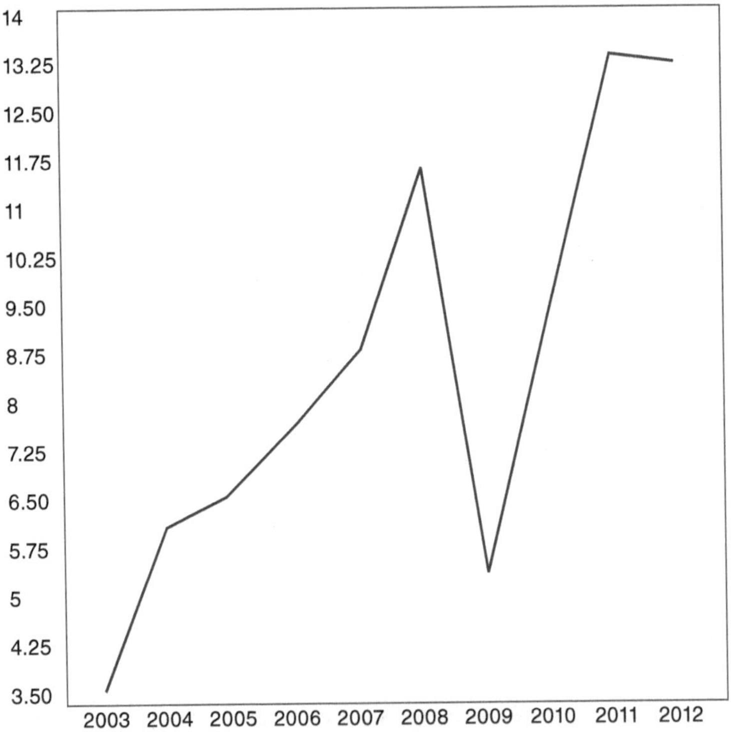

Figure 11.5 Chevron, earnings per share, 2003–2012

Source: Prepared by author from raw data at S&P Stock Reports

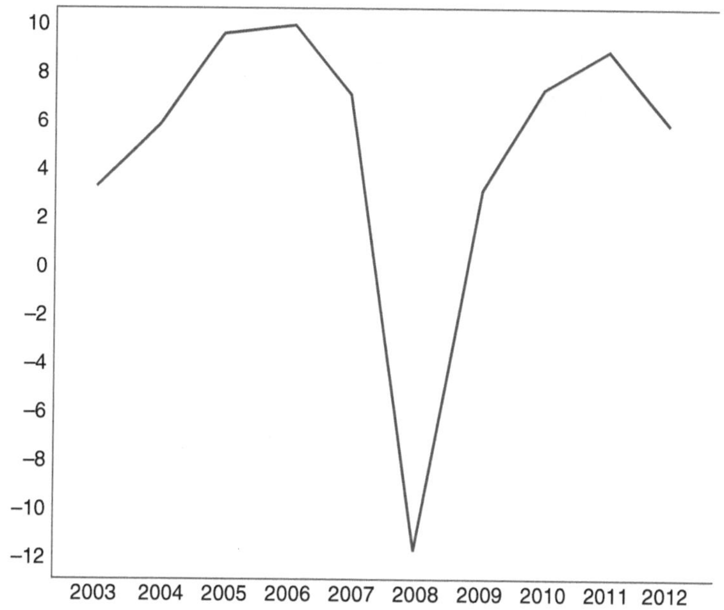

Figure 11.6 ConocoPhillips, earnings per share, 2003–2012
Source: Prepared by author from raw data at S&P Stock Reports

The comparisons between companies, notably in a check of year-to-year volatility, provide insight into the market risks for each of these three. COP is the most volatile but also pays the highest dividend; these offsetting factors have to be judged as you make a decision. However, the year-to-year EPS has to be analyzed on another level as being unique to the company itself and as one of several fundamental tests worth performing in the selection of investments and in the timing of entry and exit.

A third test, common equity, is the dollar value of common stock. On the surface, it would seem that this is a simple fundamental indicator to track. However, there are two considerations to bear in mind. First, the dollar value is changed by profits or losses because it includes retained earnings (an item that absorbs each fiscal year's profit or loss into common equity), and it also changes if and when the company issues new stock or retires existing stock.

Second, common equity is usually reported without adjustment in tangible assets. But removing intangibles results in a reduction in the dollar value of tangible common equity, or the value of common stock. For some analysts, a company-to-company comparison is more reliable on this basis, notably when the dollar values of intangible assets are quite different between those companies analyzed.

Regarding a company's long-term trend, remember to ensure that every year's definition of "common equity" is calculated on the same basis. For example, Exxon Mobil has experienced tremendous growth in common equity due to its overall profitability. Its value fell in the years 2008 and 2009, but this is to be expected given broader market conditions. XOM's common equity trend is summarized in figure 11.7.

Although all energy stocks went through the same price declines as the broader market, Chevron's common equity rose steadily. Even though its earnings per share declined, the common equity was consistent in its growth throughout the decade. If you were comparing

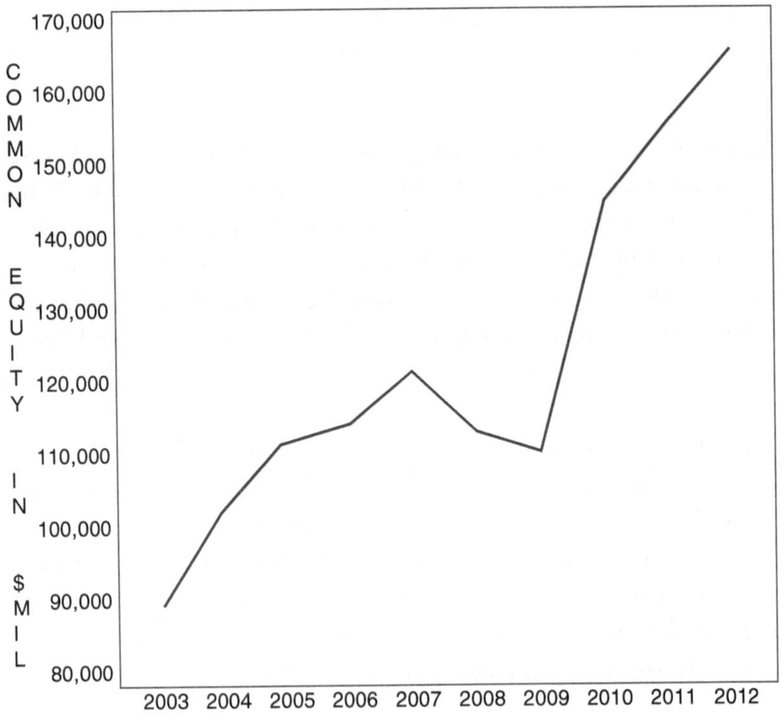

Figure 11.7 Exxon Mobil, common equity, 2003–2012
Source: Prepared by author from raw data at S&P Stock Reports

these three energy stocks, this aberration should lead to further investigation. The trend is summarized in figure 11.8.

Of the three companies, ConocoPhillips displayed the greatest volatility in its common equity. You would expect *less* volatility and not more based on the energy sector. However, revenue and profits were down in 2008 (see chapter 6), and this effect is also evident in COP's earnings per share and common equity. (Remember, common equity includes retained earnings, so a net loss in earnings will directly cause a decline in common equity.) COP's common equity for the decade is summarized in figure 11.9.

To most investors, the volatility in revenue and earnings, earnings per share, and common equity would be troubling, but two offsetting factors have to be brought into the comparison. First, COP pays a very impressive dividend far above that of its two competitors. Second, of the three companies, COP has the lowest market capitalization. This may explain the volatility in all of these indicators. A larger market cap will tend to even out the year-to-year changes, so COP's volatility might reflect the fact that its market cap is less than

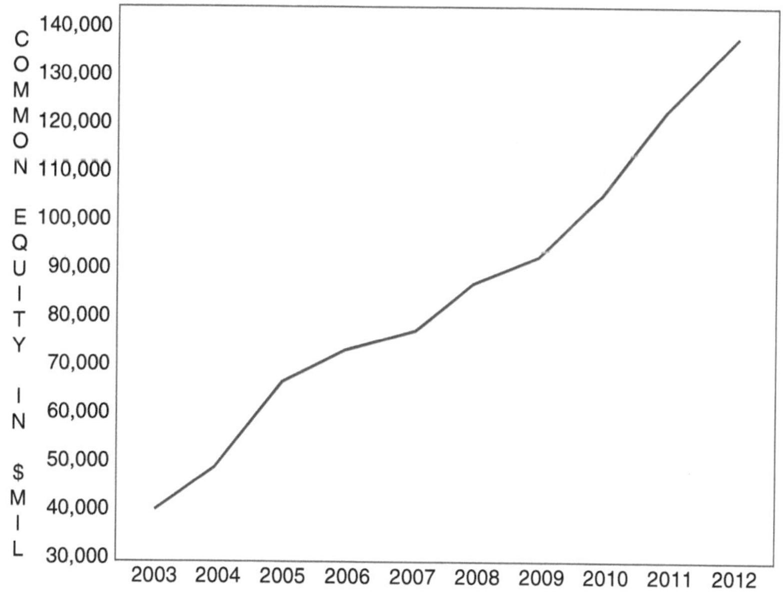

Figure 11.8 Chevron, common equity, 2003–2012
Source: Prepared by author from raw data at S&P Stock Reports

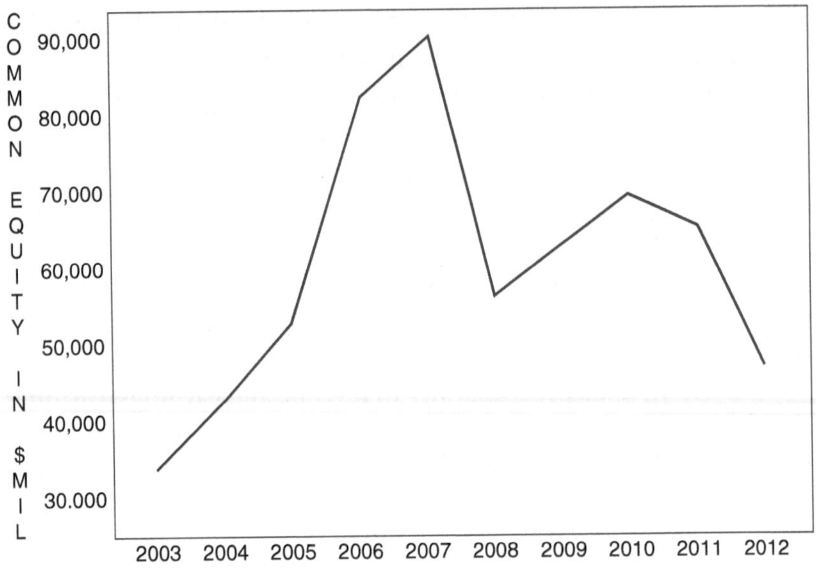

Figure 11.9 ConocoPhillips, common equity, 2003–2012
Source: Prepared by author from raw data at S&P Stock Reports

one-third that of Chevron and less than one-fifth that of Exxon Mobil.

Key point: Volatility in some indicators might be made more acceptable when the company pays a better than average dividend. You have to balance these factors against one another to make the right decision for you.

WHO IS INVOLVED?

Fundamental analysis is based largely on reported financial results. Corporations listed on an exchange publish a balance sheet listing assets, liabilities, and net worth and a statement of earnings listing revenues, costs, expenses, and profits.

Corporations and their financial departments track all of the transactions during the year and prepare preliminary statements, documentation, and files, and they interpret the accounting rules and standards according to GAAP (Generally Accepted Accounting Standards).[4]

At the end of each fiscal year (and for larger corporations, quarterly or even more often) an independent audit is performed by an outside auditing firm. These professional accounting organizations check the company's books and records and ensure that the final reports are accurate and fair. If they disagree with some of the company's transactions, they suggest changes, and once the company has agreed and made changes, the auditing firm states that in its opinion, the financial statements are fair and complete.

Corporations also have to report to regulatory agencies including the Securities and Exchange Commission and state securities agencies. These regulators ensure that the law is being followed and that corporations accurately reflect their valuation, working capital, and profit or loss in each fiscal year.

In spite of careful monitoring and reporting of financial results, abuses of the reporting standards do occur. Abuses include outright false claims, deferred reporting of expenses, early reporting of revenue, and exaggerated estimates of reserves and potential future losses. Such abuses represent a risk to all investors. Even a carefully audited financial statement is not an ironclad guarantee of accuracy or completeness. Even with the abuses, the system for reporting and auditing financial statements is of high quality in most instances. Although investors in the energy sector need to be cautious, the reliability of financial reports is very high.

JUDGING THE MARKET BASED ON THE FUNDAMENTALS

Energy stocks are subject not only to broader market forces, but also to industry-specific influences. These include geopolitical and economic forces. Investors may identify a short list of fundamental indicators based on the financial history of a company to select the strongest energy sector companies. However, it is also essential to move beyond company-specific fundamentals and analyze the sector-specific fundamentals as a means for further selection.

> **Tip:** The energy sector is unique because its fundamentals occur on two levels: on the level of the company and that of the entire sector.

The market overall as well as individual sectors and companies all rely on fundamental trends to assess their strengths and weaknesses. All sectors go through market cycles based on the economy as well as popularity among investors. In addition, the energy sector has a unique attribute: The perceptions of "value" and "risk" are often false but may still be more important than the realities. These have been explored in detail earlier in this chapter and in preceding chapters. A review follows here:

1. *Energy scarcity affects prices.* The scarcity of energy occurs even though new reserves and extraction methods are developed each year. Today, the United States has ample reserves of natural gas and oil, but energy is widely believed to be scarce and about to run out.
2. *Environmental concerns dominate.* The environmental lobby has done an exceptional job of convincing many citizens that oil damages the ozone layer and air and water quality, that alternative energy is cheaper, and that the answers to energy supply questions are simple. For example, ethanol has been promoted as the answer to energy problems. The US government mandated that a specific level of corn production must go into ethanol, but this has had unintended consequences:

> Since 2005, the US government has mandated that gasoline contain ethanol, almost all of it derived from corn. The policy, ostensibly aimed at reducing the country's dependence on foreign oil and at improving the environment, has been a bonanza for farmers. The acreage planted with corn soared by a fourth after Congress passed the Energy Independence and Security Act of 2007, which required that gasoline producers blend 15 billion gallons of ethanol into the nation's gasoline supply by 2015.
>
> Now the drought of 2012, the worst in more than 50 years, is making clear the downside of a policy that leads the United States to devote 40 % of its corn harvest to fuel production. With this year's crop expected to be the smallest in six years, corn prices have jumped 60 percent

since June. The ethanol requirements are aggravating the rise in food costs and spreading it to the price of gasoline, which is up almost 40 cents a gallon since the start of July.[5]

3. *Geopolitical unrest affects domestic prices in spite of reduced reliance on imports.* With only 13% of US oil coming from the Persian Gulf and vast reverses of gas and oil at home, it is entirely possible for the United States to become self-sufficient in just a few years.

   Discussions of US strategic interests in the Middle East tend to focus on generalizations about broad strategic interests, oil, trade, friendship, and peace negotiations and then show concern over all the "usual suspects" like instability, arms sales, rogue states, proliferation, and terrorism.[6]

In fact, US interests in the region are more complex and include a balance of power with the Russians and preventing attacks on the United States by any number of enemies in the region. In addition, the United States is concerned with ensuring energy security not only for itself but also for its allies and with supporting Israel, its only democratic ally in the Middle East as well as with preventing Iran and other radical countries from getting nuclear weapons.[7]

4. *Domestic partisan politics affect energy decisions.* The question of whether or not to expand domestic drilling (via pipeline, offshore drilling, in the Northern slope of Alaska and in oil sand and shale reserves) is not decided by market forces as much as by political party affiliation. The support of the environmental lobby has a direct influence on those in office, and votes tend to follow party lines to remain in favor with groups offering support for a particular position. Were these restrictions removed, the vast reserves could be tapped. However, from the investment point of view, would this help or hurt? A more plentiful supply often means lower prices and reduced demand. From a purely economic viewpoint, then, opening up all of the sources of energy might be good for fuel consumers but not so good for investors.

THE LONG-TERM TREND

In the study of fundamentals, the longer a trend is analyzed, the better the information. In studying the energy sector, the study of a short list of fundamental indicators over 10 years reveals strengths and weaknesses of companies and reveals important trends. This has been demonstrated in this chapter as well as in chapter 6. The various trends make greater sense to those taking a long-term perspective, and a majority of company-specific fundamental indicators make sense only when viewed within the context of a long-term trend.

> **Tip:** When working with the fundamentals, the longer the trend you analyze, the better able you are to see what it reveals—not only about the past, but about the future as well.

The many fundamental indicators do not require an accounting background to understand. In fact, trends are a method for making the numerical data visual. Everyone can spot a positive or negative trend and a reversal in direction. Charting ratios and numerical values is not difficult, and with modern technology, raw data can be fed into programs that create charts (for example, in MSN Excel or Adobe Illustrator).

Charting of ratios can be performed in several ways, including:

1. *Dollar values.* The most basic of charts depicts dollar values. When dealing with millions or billions of dollars for large cap companies, it is common to exclude six zeroes, so that the results are in the billions. Thus, $59,000 actually is a shorthand expression of $59,000,000,000 or $59 billion. Charting dollar values makes trends easy to spot, whereas a column of numbers cannot be interpreted as easily.
2. *Ratios developed from dollar values.* When you are dealing with two factors, developing and then plotting a ratio makes the most sense. For example, the debt ratio is actually a percentage, representing debt as a percentage of total capitalization. The easiest form of charting the debt ratio is to track a 10-year history of the percentage.

3. *Percentage of change from one period to another.* Another version is to plot not the period's results but the percentage of change from one period to another. This may seem like a great statistical technique, but it can also distort the results. A value of 100 followed by 200 is a 100% increase, and if the following period is then back down to 100, this is a 50% decrease. Using this technique is appropriate in some situations, but in others it clouds the trend. Most fundamental analysis is based on comparisons between periods rather than percentage changes in each period.
4. *Charting of several results side by side.* Many trends are most revealing when two trends are reported together. For example, including revenue and net earnings on the same chart reveals how well profits track the revenue trend; this is useful information. For example, if the earnings trend is slower than the revenue trend, it could mean management is not controlling expenses and that the margins are falling even though the numbers are rising.

USING FUNDAMENTAL INDICATORS

Energy investors, like everyone else, need to create a list of valuable but limited indicators to track. These should include at the very least the trend in revenue and earnings, P/E ratio, dividend yield, and the debt ratio.

Many other fundamental indicators unique to each company may improve insight into the trends, notably when dealing with a short list involving prices and earnings (PE ratio), dividend yield and payout ratio, revenue and earnings, and debt ratio. These are the basics; you will find many indicators beyond them, but they are not as useful.

Key point: Keep your list of indicators fairly short. Adding more tests does not improve your information, and the longer the list, the more difficult it is to get to the decision point.

The danger in analyzing too many indicators is twofold. First, when you gather too much information, it becomes increasingly difficult to make a trade decision; limiting yourself to only a few strong fundamentals is more effective. Second, some fundamental indicators are not only obscure but provide little if any useful information. A useful indicator should reveal something of value and answer your questions, such as:

1. *Is this company profitable?* This is the most basic question every fundamental investor needs to ask. There might be circumstances where you will gladly invest in a company that is not yet profitable as long as you are confident that the future looks bright. But as a general rule, it makes little sense to invest capital in a company that is not turning a profit.
2. *Are revenues and profits improving or leveling out?* The growth in the dollar value of revenues and earnings should be steady, but you have to expect every trend to level out eventually. The revealing test is to compare not the dollar amount but the net return. In an ideal situation, even as revenue growth begins to slow, the net return should remain the same. If a company has been reporting net return of 9% every year, you want to see 9% even when revenues are not growing as rapidly as in the past. If net return begins declining, it could be a signal that the company is losing its competitive edge. The worst situation is when revenues are growing but profits are eroding. That might be an early warning that it's time to move your money elsewhere.
3. *Is management controlling debt?* You hope to see a company's debt ratio remain steady, lower than that of its competitors, and perhaps even falling during periods of growth. If a company's debt ratio is growing, it is a troubling signal. In a healthy situation, a company's debt ratio should be at or below that of its competitors. A rising debt ratio is a danger signal. It means that future earnings have to go more and more to interest and repayment of the debt, which means there will be less profit left over to pay dividends and fund operations.
4. *Is the current price realistic in view of recent results?* This is more technical than fundamental, but it relates back to how well management handles its fundamentals—revenue and profits. The price of stock is a reflection of how well management is able to create growth and profits over the long term. The P/E ratio is reported as a multiple (the number of years of growth

in earnings per share represented by today's price; for example, a P/E of 17 reveals that the current price per share is equal to 17 years of net profits at currently reported levels). When you see a price of $35 per share and EPS of $2.00, that means the P/E is 17.5 ($35 ÷ 2). That's in the ballpark of a midrange P/E (usually thought of as between 10 and 25). But if the P/E is 60, that means the current price is equal to 60 years of earnings, and that is quite high. This reflects popularity in the market, but it may also indicate an overpriced stock.

The next chapter moves to a related topic, technical analysis. In the energy market, as with most sectors, a combination of fundamental and technical indicators creates an effective and powerful analytical means for timing your trades.

CHAPTER 12

TECHNICAL ANALYSIS OF THE ENERGY MARKET

> I can stand brute force, but brute reason is quite unbearable. There is something unfair about its use. It is hitting below the intellect.
>
> Oscar Wilde, *The Picture of Dorian Gray*, 1891

TECHNICAL ANALYSIS HAS MANY VARIATIONS. MOST TRADERS associate this form of analysis with price movement and trends. However, in addition to price indicators, technical analysis also includes the study of volume and momentum. Energy investors and traders can make good use of all of these tools to improve the timing of entry and exit into stock positions. By timing these decisions based on reversal indicators found on charts, you will be able to spot coming changes early and to take advantage of what the indicators reveal.

The idea is that with the study of many technical indicators, traders may identify reversal signals. If these are confirmed by another technical signal, chances of good timing for entry and exit are vastly improved. Technical analysis is further divided into Western and Eastern. Western technical analysis includes the concepts of support and resistance, price gaps, patterns such as head and shoulders or double tops or bottoms, wedges or triangles, flags or pennants, and many more. Eastern technical analysis is based on analysis of candlesticks in single sessions, double sessions, or complex (three or more) sessions.

Key point: Western and Eastern technical analysis work well together and may confirm one another, especially for reversal signs.

While the detailed study of the many technical signals aids in identifying entry and exit timing and thus improves frequency of profitable trades, energy traders will find that the use of technical analysis along with fundamental analysis creates the most reliable system. The premise of using both is that as a first step, you identify companies whose quality is high enough to justify investing; after buying shares in these companies, technical indicators are used to test market risk and to identify price reversals.

What do traders do upon spotting reversal? Many close equity positions in the belief that the signals—especially bearish reversal signals—are timing tools and that locating a bullish reversal later indicates that it is time to get back into the market.

There are a few problems with this approach. First, by selling stock, you give up any dividends earned during the period stock is not held. Second, even with the best indicators, reversal and confirmation are not 100% reliable. So you might sell at what appears to be a price top and with a strong reversal signal, only to then see the stock price continue to rise.

A more prudent approach is to first locate high-quality value investments that also yield a higher than average dividend. Hold these shares but track the technical indicators as well. Use options to swing trade the position, so that you can take profit and hedge against risks at the same time. Options are cheap compared to the potential cost of ill-timed stock sales. Instead of taking profits by disposing of stock, long calls and puts are very cheap and can be used to take advantage of reversals (buying puts when bearish reversals are spotted and buying calls when bullish reversals show up).

Options allow you to take profits and to take advantage of short-term swings without having to dispose of stock you would rather keep over the long term. Your timing will be off some of the time, and here options are a better choice than disposing of shares. Because long options are so cheap, you can afford to have losses mixed in with profits.

The two-part strategy is to hold stock in your portfolio and earn income from dividends and to use long options to swing trade against those positions, increasing current income while reducing downside risk. If you consider identification of value stocks based on fundamentals, it means you want to hold them for the long term and earn dividends; you believe that these are low risks as well. At the same time, a technical swing trading strategy enhances income and lets you manage market risk. This is where low-priced options can be the most valuable. The combination of a solid fundamental stock portfolio with risk reduction achieved through the hedging and profit potential of options is a popular strategy today. Identify strong investment candidates based on fundamental analysis and trade the short-term swings based on technical analysis.

> **Tip:** Study options to find an exceptional combination of risk reduction and increased current income. Options can be high-risk investments—*or* they can be used in very conservative strategies.

How cheap are options you would use in this two-part strategy? Table 12.1 summarizes option values for XOM as of the close of trading on May 23, 2013.

The farther out the expiration date is, the higher the percentage (option premium divided by current price per share) runs. This is a reflection of time value premium. None of these prices exceed 7.5%, of stock prices, so that you can control 100 shares of stock for less than 7.5% of the value.

These relationships are approximately the same for all three energy companies studied in this book. For example, table 12.2 shows the same data set for Chevron.

Table 12.1 Option premium, Exxon Mobil, May 23, 2013

Stock price $91.80 Month	**Strike**	**calls**	**%**	**puts**	**%**
JUN	90	2.72	3.0%	0.91	1.0%
	92.50	1.13	1.2	1.80	2.0
	95	0.33	0.4	3.31	3.6
JUL	90	3.30	3.6%	1.43	1.6%
	92.50	1.75	1.9	2.40	2.6
	95	0.76	0.8	3.85	4.2
AUG	90	3.80	4.1%	2.15	2.3%
	92.50	2.28	2.5	3.20	3.5
	95	1.20	1.3	4.50	4.9

Source: Charles Schwab & Co., options listings, May 23, 2013.

In this case, the percentage of stock price represented by option premium was much higher than for Exxon Mobil. This occurred because the distance between strikes was five points; that is, more contracts were in the money, which means higher intrinsic value. In comparison, Exxon Mobil's strikes were only 2.5 points apart. This demonstrates that option values are a combination of time and the proximity between current stock price and the option's strike. Even here, for 7.5% or less of the stock's price you can control 100 shares of stock using options.

The third stock, ConocoPhillips, had a similar relationship in the options on the same day, as shown in table 12.3.

Table 12.2 Option premium, Chevron, May 23, 2013

Stock price $124.60 Month	**Strike**	**calls**	**%**	**puts**	**%**
JUN	120	6.05	4.9%	0.66	0.5%
	125	2.36	1.9	1.93	1.5
	130	0.49	0.4	5.05	4.1
JUL	120	6.68	5.4%	1.30	1.0%
	125	3.26	2.6	2.77	2.2
	130	1.19	1.0	6.51	5.2
AUG	120	7.40	5.9%	2.25	1.8%
	125	4.04	3.2	4.05	3.3
	130	1.77	1.4	7.20	5.8

Source: Charles Schwab & Co., options listings, May 23, 2013.

Table 12.3 Option premium, ConocoPhillips, May 23, 2013

Month	**Stock price $62.79 Strike**	**calls**	**%**	**puts**	**%**
JUN	60	3.25	5.2%	0.43	0.7%
	62.50	1.42	2.3	1.09	1.7
	65	0.33	0.5	2.62	4.2
JUN	60	3.50	5.6%	0.90	1.4%
	62.50	1.72	2.7	1.86	3.0
	65	0.75	1.2	3.35	4.3
JUN	60	3.55	5.7%	1.40	2.2%
	62.50	2.05	3.3	2.35	3.7
	65	0.99	1.6	4.00	6.4

Source: Charles Schwab & Co., options listings, May 23, 2013.

In this case, the combination of time and proximity defined option premium once again. For a cost below 6.5% of the cost of 100 shares, you can control 100 shares of stock using options.

TYPES OF TECHNICAL INDICATORS

Technical analysts study price patterns to identify and anticipate when price direction is likely to reverse. This field of study also includes observation of trading volume and tracking of momentum oscillators, which show the speed and strength of a trend. The following are examples of typical indicators in each of these three areas (price, volume, and momentum).

PRICE INDICATORS

Bollinger Bands are a tracking indicator for measuring volatility in a stock's price. Volatility is measured statistically by calculating *standard deviation*,[1] the dispersion of prices away from an average or the difference between actual and average price. The higher the standard deviation is, the greater the volatility. Bollinger Bands calculate two moving averages based on standard deviation and include a middle simple average of the past 20 sessions.

Key point: Bollinger Bands identify volatility in a visual manner by displaying in the upper and lower bands two standard deviations away from the middle. The wider the bands appear, the higher the volatility.

This is the basis of Bollinger Bands. This indicator makes volatility visual; as the latter increases, the bands move away from the center and become wider; as volatility decreases, the bands narrow. The upper and lower bands are usually calculated at two standard deviations above and below the simple moving average. Comparing the simple moving average (the baseline or average) to the higher and lower bands provides a visual representation of volatility.[2]

Bollinger Bands have been the topic of keen interest among traders since the 1980s when John Bollinger first introduced the idea. As a means for identifying the range of likely price movement and labeling volatility as either high or low within that range, the bands have been studied extensively and have been shown to be very reliable. Approximately 88% of a security's price range remains within the Bollinger Bands during the periods being studied.[3]

Some traders interpret Bollinger Bands to signal entry or exit. One popular idea is that when price touches the lower band, this is a buy signal, and when price rises to touch the middle band, this is a sell signal. A price breakout above the upper band could be interpreted as strongly bullish, and a breakout below the lower band is the opposite, a strongly bearish signal. Options traders see expansion of the band's width as a signal to sell existing long positions or to open short positions based on the heightened volatility. The idea is that high volatility is normally a very momentary change, accompanied by higher than typical option premium. Once the band's width returns to a more narrow normal state, any newly opened trades are closed, or the status of the range reverts to its midrange; traders then wait for another move in price toward the upper or lower band levels.

Bollinger Bands serve as a strong confirmation indicator, notably when other price or momentum signals indicate reversal of the existing price direction. As an example, the chart of Exxon Mobil is shown in figure 12.1 with an overlay of Bollinger Bands.

The band's width was fairly small for the first half of the chart, and price remained in a narrow trading range. However, at the beginning of April the bands widened, and on April 15 price dropped below the lower band. One interpretation of this was that a reversal was about to occur. This was confirmed two sessions later with the exceptionally long lower shadow, signaling lost momentum among sellers. Price then moved strongly upward in a series of gapping sessions.

Bollinger Bands provided a structure to this price pattern and enabled traders to identify a likely turnaround in the trend. But confirmation was essential before investors could act on the information. Some of the candlestick signals toward the end of April looked like bearish reversal, but the Bollinger Bands contradicted that and did not reveal any likelihood of a price change. Toward the end of the period charted, the lower band increased its distance from price, which is open to interpretation. Comparing the first half of the period (when the lower band tracked very closely to price) to the second half (when the gap between price levels and the band grew considerably) should direct traders to look for signals of a possible reversal yet to come.

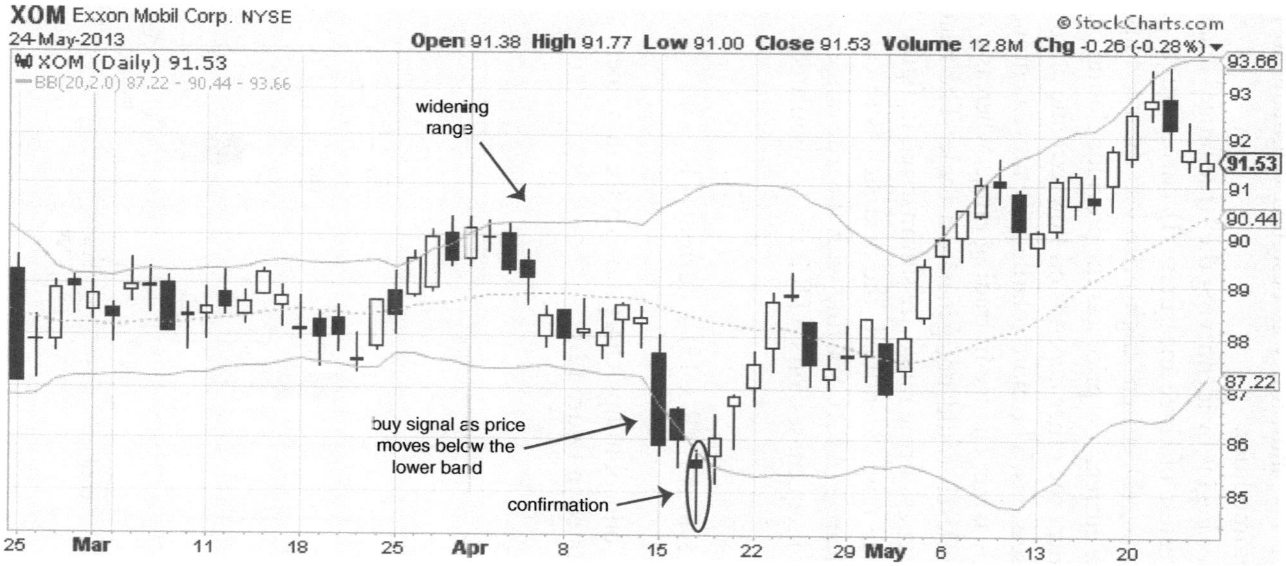

Figure 12.1 Exxon Mobil, Bollinger Bands
Source: Chart courtesy of StockCharts.com

Average True Range (ATR) is another indicator used to measure volatility in price. J. Welles Wilder developed ATR in a book published in 1978. The indicator was designed originally to measure volatility in commodities.[4]

The indicator begins by identifying the "true range" of a security. This is the greatest of three measures: (1) current high price less current low price; (2) current high price less the previous closing price; or (3) the current low price less the previous closing price. The choice of one over the other depends on how extreme price changes were between sessions. The more volatile price movement indicates a correspondingly larger change in trading range, whereas a relatively small volatility level would not generate a trade. The relationship and the three types of true range are summarized in figure 12.2.

With the true range identified, the next step is to calculate the Average True Range (ATR). Starting with the true range in the first session, each subsequent session's ATR is the accumulated true range of 14 sessions:

ATR = ((previous ATR x 13) + current TR) ÷ 14

For example, ConocoPhillips' chart shows price movement as well as ATR, as illustrated in figure 12.3.

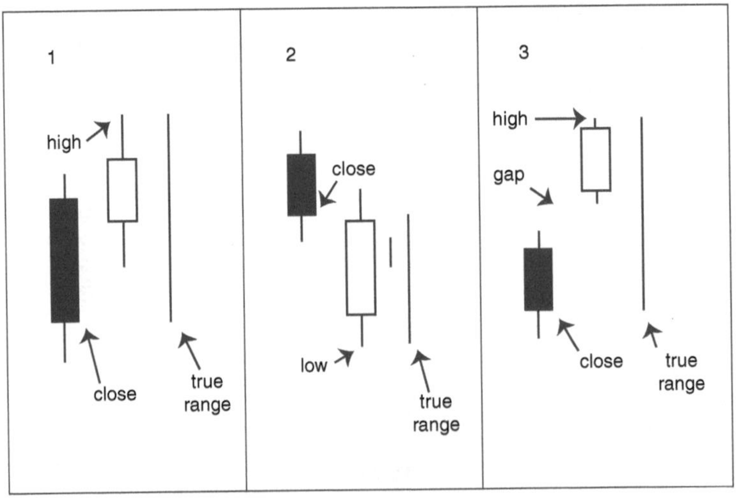

Figure 12.2 True Range
Source: Created by the author

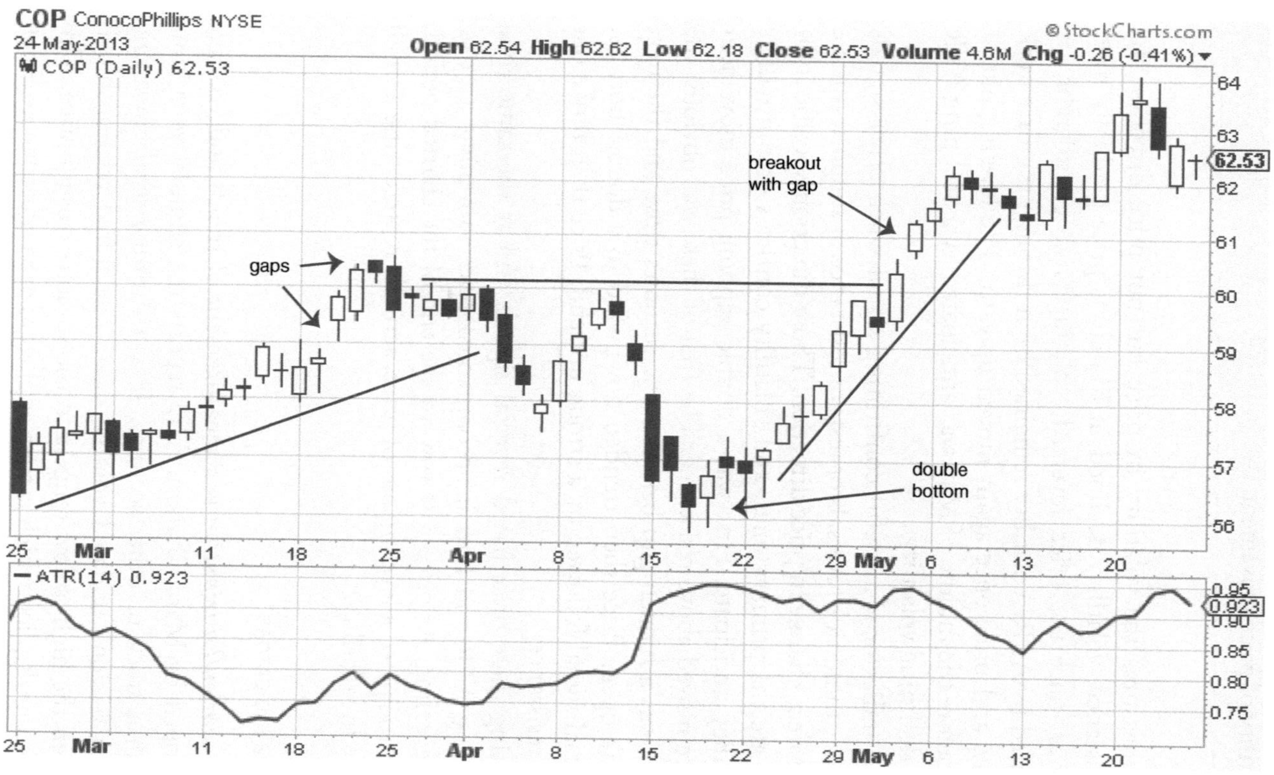

Figure 12.3 ConocoPhillips, Average True Range
Source: Chart courtesy of StockCharts.com

> **Tip:** Indicators like ATR are simplified, but they help put price into context; in the case of ATR, the true range reveals the current volatility of price movement.

ATR measures the volatility in price but not price direction. In this instance, volatility declined in March as price gradually rose. However, once the price reversed, volatility rose considerably as measured by ATR. Volatility remained high through the first week of May, and price once again moved upward. The leveling out of price before the middle of May was anticipated by a declining ATR (thus, declining volatility).

This indicator is a useful confirming signal when other signals (such as candlestick reversals or tests of resistance or support) initiate an apparent change in price direction. Those direction-specific indicators are the primary ones, and volatility confirms or contradicts what they anticipate. First was a double gapping price movement as price moved above resistance. The first of these was obvious; the second was a hidden gap between one session's close and the next session's open. Price then trended downward to a double bottom, a reversal signal followed by a strong uptrend. Price then broke out above resistance.

Noting these price-based reversal signals and at the same time observing ATR and how it accompanied price movement can be very revealing. As price rose toward the end of the charted period, ATR anticipated a leveling out of price, actually representing a small decline in volatility.

VOLUME INDICATORS

Accumulation/Distribution is a measurement of trends and volume. Developed by Marc Chaikin, A/D reports the cumulative volume flow in a stock or other security. It is a running total consisting of each session's updated volume data. The initial relationship is between the high-to-low range and the session's closing price. This creates a multiplier, which is then applied to each subsequent session to develop the A/D line. The A/D line confirms a current trend or anticipates reversal as the volume signal begins to weaken. This is based on the observed relationship between price strength or weakness and changes in levels of volume.

Calculation of A/D is done in three steps. First, the multiplier is computed; next, this value is multiplied by volume. And finally, a running total forms the actual A/D line:

1. Money Flow Multiplier = ((Close – Low) – (High – Close)) ÷ (High – Low)
2. Money Flow Volume = Money Flow Multiplier x Volume for the Session
3. A/D = Previous A/D + Money Flow Volume

This series of calculations creates the running A/D, which ranges between +1 and -1. It will be a positive value when the session's close is located in the upper half of the range of high to low and negative when in the lower half. An upward-trending A/D is bullish, and a downward-trending A/D is bearish. The A/D summarizes the strength of buying versus selling volume.

Key point: Indicators should simplify and clarify. The A/D line, for example, tracks bullish or bearish trends with a single line; this makes changes in trends easy to spot.

In the chart of Chevron, the A/D line moves into positive territory and remains there at the beginning of May, as shown in figure 12.4.

Throughout the first half of the chart, support held. Then price dipped below but—as it often occurs—then reversed and moved upward strongly. By itself, this price movement was difficult to read. However, with the A/D line tracking the same period, a short-term rise into positive territory occurred right before the reversal on April 11 and 12, and then as price began moving upwards in the beginning of May, the A/D line rose and moved into positive territory and remained there until the end of the month. When A/D moves in the opposite direction of price, it may signal a coming price reversal, but this did not occur as price moved upward again through May.

Chaikin Money Flow (CMF) is an outgrowth of A/D. Also developed by Marc Chaikin, CMF forms the basis for A/D; however,

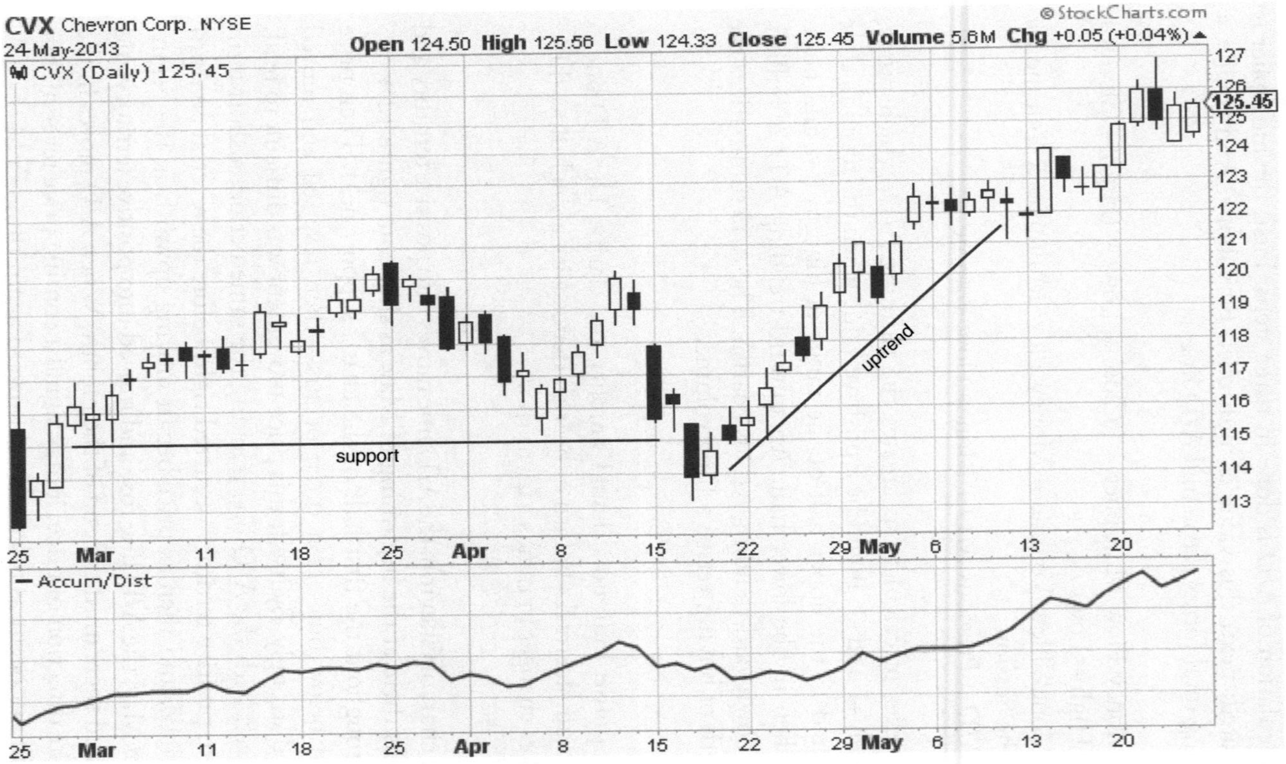

Figure 12.4 Chevron, Accumulation/Distribution
Source: Chart courtesy of StockCharts.com

instead of a cumulative value between +1 and -1 as with A/D, CMF represents the sum of money flow over a specific period of sessions, typically 20. That is, the indicator moves above or below the "zero" line based on the average of money flow over 20 sessions. Traders use CMF as a signal of change as the level moves from above to below or below to above. Most of the time, CMF will fall between +0.5 and -0.5. Any move beyond these levels would be extreme and would occur only after 20 sessions of closing at the same level to reach the extreme levels.

To calculate CMF, three steps are required. This is very similar to the A/D formula, but the distinction is that a fixed period is used to average the outcome, rather than adding the result to the prior A/D. For CMF, first compute the money flow multiplier; next calculate money flow volume by multiplying the multiplier by volume for the session and, finally, divide the 20-session sum of money flow volume by the 20-day sum of total volume:

1. Money flow multiplier = ((close – low) – (high – close)) ÷ (high – low)
2. Money flow volume = money flow multiplier x volume for the session
3. CMF = 20-session sum, money flow volume ÷ 20-session sum, volume

CMF will be positive (bullish) when the session closes in the upper half of the high-to-low range or negative (bearish) when the close occurs in the lower half. Putting this another way, CMF is at +1 when the session closes at the high, and it is at -1 when the close is at the session's low. These values represent the extremes of price range related to volume. As an example, CMF is shown on the chart of ConocoPhillips in figure 12.5.

Note the trend lines drawn on the chart in relation to CMF. The only time CMF dips into bearish territory is at the double bottom. In this instance, CMF contradicted the apparent signal, with the price dip most likely to be the turning point (double bottom), and this in fact proved to be the case. Price moved strongly upward immediately after this. CMF had momentarily given out a bearish signal, but it quickly followed the price trend and returned to positive territory.

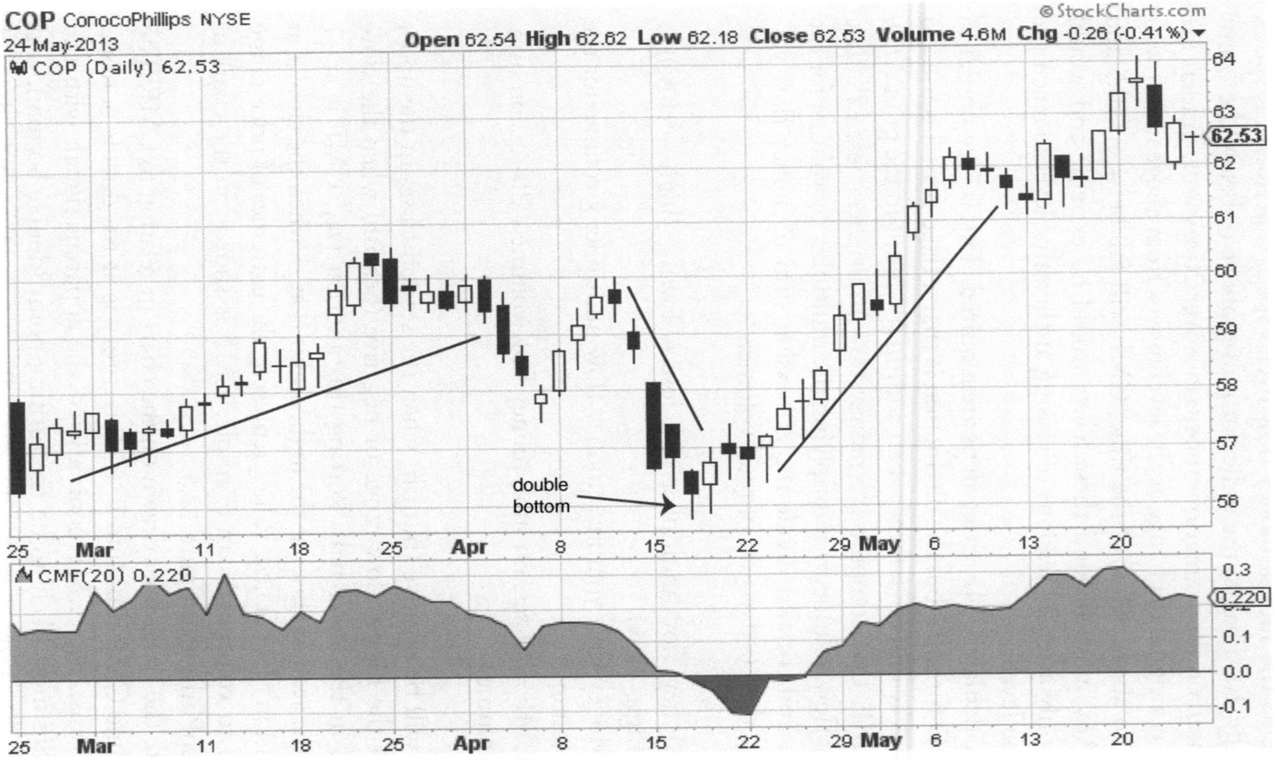

Figure 12.5 ConocoPhillips, Chaikin Money Flow
Source: Chart courtesy of StockCharts.com

Tip: No indicator will work 100% of the time. No matter how strong a signal looks, always act only upon finding equally strong confirmation.

On-balance volume (OBV) is a cumulative indicator, like the A/D line. Invented by Joseph Granville, OBV tracks positive and negative volume flow and may be most useful when its signal contradicts what price indicators forecast. When a session's price moves up in a session, the day's volume is added to the cumulative OBV, and when the price closes down, the volume is subtracted.[5]

The flaw in this method is that whether prices moves a small fraction of one point or moves by a double-digit move in the same direction, volume is treated in the same manner. That is, distortions can occur. For example, price might close slightly up over five consecutive sessions, followed by a very large downward-moving session. In this outcome, the first five days' volumes are added, and the final session's volume is subtracted. Depending on the volume level, this treatment can easily distort the outcome of OBV calculation and mislead traders. Like many other indicators, OBV is only one test of volume and its effect on trends, and it should be used as a means of confirmation rather than as a leading reversal signal.

The value of OBV is its measure of the strength in buying or selling pressure and how this pressure affects price. The indicator is a cumulative net between positive and negative volume (positive means the session closes higher than the prior session, and negative means the session closed lower). To calculate follow these steps:

Positive session: prior OBV + current session's volume
Negative session: prior OBV – current session's volume
Session's close is equal to prior session: prior OBV (no change)

OBV is based on the belief that volume precedes and predicts changes in price direction. Thus, if volume pressure is positive, then price levels will also become positive. Chartists tend to use OBV to confirm or anticipate price trends and to see whether OBV confirms the direction in price. OBV nay also anticipate breakouts above resistance or below support. When OBV changes direction at points where price is near these important lines, it could foreshadow a breakout and the development of a new trading range.

OBV may also foreshadow a change in price direction when it diverges from price. If the price is in an uptrend but OBV turns negative, this could serve as a signal that price may reverse very soon. Traders would then look for confirmation in other indicators related to price or momentum. Divergences are likely to be the strongest when price is near resistance or support.

Key point: Resistance and support serve as crucial parts of interpretation. Most technical signals will be at their strongest and most reliable when price is close to resistance or support and may either break through or reverse course.

These distinctions (price anticipation or divergence) are less common than trend confirmation seen with OBV. When a trend is underway, traders will tend to believe it will continue when confirmed by indicators such as OBV. The chart of Chevron in figure 12.6 includes the OBV line.

OBV tracks the short-term price trends extremely closely through the first half of the chart. However, OBV outpaces the price as the uptrend begins. That uptrend is marked at the bullish harami and confirmed with the large number of upward gaps. But at the same time OBV continues to move further upward, showing that the buying pressure was quite strong and continuing. It was not until May 13 that OBV declined back to the range of price levels, showing that the strong uptrend was ending. This did not mean the uptrend was over, but it did signal that the strength of buying pressure had declined.

Traders recognize the evolution of OBV relative to the price trends and take a cautionary stance, perhaps looking for a reversal signal to occur next and an adjustment to the downside to then follow. If nothing is found, then the price-specific signals would remain in effect until conditions change. It appears that the very sharp and strong uptrend from April 20 through May 6 developed into a slower uptrend from May 6 to May 31—not a reversal, just a more gradual version of the uptrend.

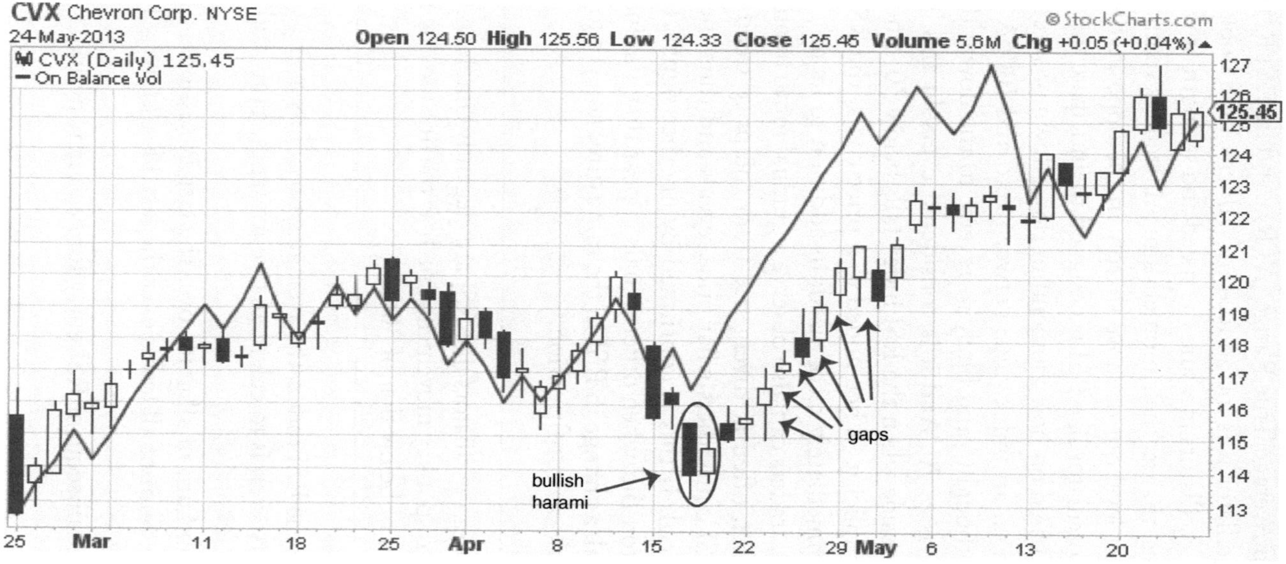

Figure 12.6 Chevron, On Balance Volume
Source: Chart courtesy of StockCharts.com

MOMENTUM INDICATORS

Relative Strength Index (RSI) was developed by J. Welles Wilder, who also introduced Average True Range (ATR). And like ATR, the RSI momentum oscillator was introduced in Wilder's 1978 book.[6]

Momentum oscillators measure the speed and strength or weakness in price changes, that is, their momentum. All trends eventually lose steam and either level out or move sharply in the opposite direction. A momentum oscillator like RSI is designed to anticipate these directional changes based on how momentum slows down or speeds up.

RSI is based on an index value between zero and 100. A stock is treated as "overbought" and likely to reverse to the downside when this index surpasses 70. And it is considered oversold and likely to reverse to the upside when the RSI index falls below 30. The appeal of RSI is its simplicity. When added to a price chart, the RSI status is easily and immediately recognized; in addition to its move above 70 or below 30, traders can also observe a gradual move in one direction or the other, often at divergence from the price trend. When that occurs, RSI contradicts what price reveals and thus anticipates a coming reversal. If this is confirmed by another signal, RSI is a leading indicator. But when other signals appear first and indicate a coming reversal, a change in RSI in the same direction is a following signal that works as confirmation.[7]

> **Tip:** Momentum oscillators like RSI are elegantly simple, making it quite easy to spot key movements, such as the line moving above the level of 70 or below that of 30.

To calculate the RSI index, the average gain or loss over 14 periods is used. The first step is to calculate two averages, gains and losses:

Average gain: Sum of all gains over the last 14 sessions ÷ 14
Average loss: Sum of all losses over the last 14 sessions ÷ 14

Next, calculate subsequent RSI entry for each new period:

Average gain: ((Prior average gain x 13) + current gain) ÷ 14
Average loss: ((Prior average loss x 13) + current loss) ÷ 14

From these calculations, the overall net average is calculated:

Relative Strength (RS): Average gain – Average loss

With these averages in hand, RSI is calculated with the following formula:

RSI = 100 – ((100 ÷ (1 + RS)))

The calculation can be modified in two ways. The use of 14 periods can be changed. Using longer periods tends to smooth out the RSI line, and using shorter ones tends to accelerate movement. Second, the use of the index values 70 and 30 can be changed; for example, one modified version of RSI is based on the use of 80 and 20, a more conservative definition of overbought or oversold. The decision to change either the number of periods or the overbought and oversold levels may rely on the volatility of the security; a higher than average level of price volatility may justify longer averages and higher as well as lower overbought/oversold values. An example of RSI is found on the chart of ConocoPhillips shown in figure 12.7.

This chart provides an example of how RSI confirms reversal signals found in candlestick formations. The first time RSI reached the all-important level of 70, price moved above resistance and immediately formed a bearish meeting lines signal. All candlestick reversals are strongest when they occur at resistance or support, especially when price moves through and the candlestick forms a reversal back to the established trading range. This occurred on the chart and was confirmed by RSI slightly above 70.

The bearish piecing lines occurred while RSI remained in the midrange; however, price did retreat as the candlestick signal predicted. But the strongest reversal on the chart was the bearish abandoned baby, which occurred at the same time that confirmation was seen in RSI moving above 70 yet again. Even though the move was slight, it was adequate confirmation for a signal as strong as the bearish abandoned baby.

Key point: RSI is revealing and often foreshadows coming price changes. But you need confirmation. Indicators found in candlesticks, for example, are strong forms of confirmation that add to your confidence in timing entry or exit.

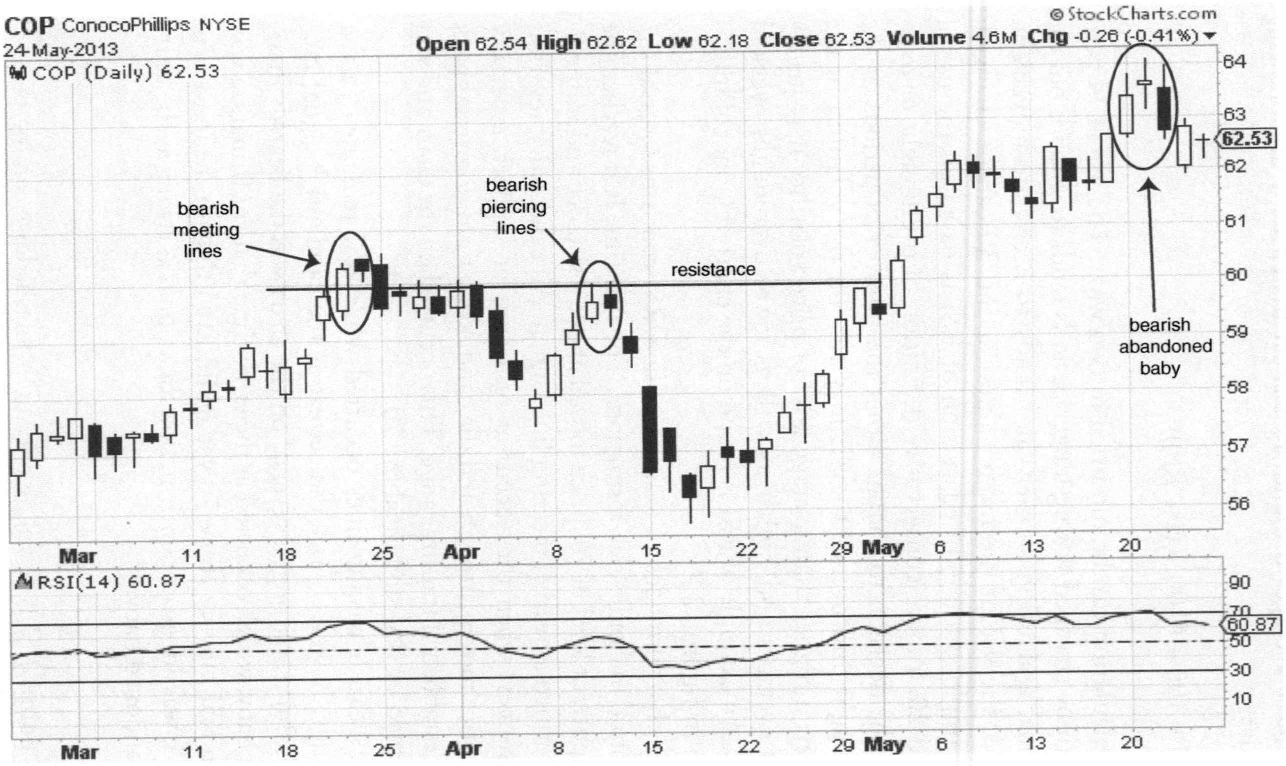

Figure 12.7 ConocoPhillips, Relative Strength Index
Source: Chart courtesy of StockCharts.com

RSI divergence is another very crucial development. A bullish divergence consists of RSI forming to a new high while price declines, perhaps as drastically as moving to a new low. A bearish divergence is the opposite: a new low for RSI accompanied by a new high in price. Divergence can be misleading and serves as a false signal when the price trend is exceptionally strong; divergence by itself should not dominate the decision whether to enter or exit a trade. However, when confirmed by other signals, such as tests of resistance or support, price gaps, or double tops or bottoms, divergence confirms the change expected in price. For example, if price gaps upward to form a double top while RSI dips to a low, this probably signals a coming reversal and downtrend.

Failure swings occur for many reasons, and what appears to be divergence in RSI might only be a symptom of retracement and temporary trend adjustment. This points out the need to have any signal accompanied by strong confirmation. RSI is popular because of its simplicity and reliability, but like all signals, it can mislead as well. You can avoid failure swings by requiring confirmation before acting. The failure swing may cause traders to exit too soon or even to enter a position at the worst time; these unfortunate outcomes are lessened when the signal occurs but is not confirmed. Failing confirmation, a wise trader will not react to RSI. An unconfirmed signal is typical of retracements, and they require caution.

The stochastic oscillator is a momentum indicator reporting the proximity of closing prices for each session to the high-to-low range over a specified number of sessions, usually 14 consecutive sessions.

The stochastic oscillator does not track price but the speed of movement in price. George C. Lane stated in an interview that the stochastic oscillator "doesn't follow price, it doesn't follow volume or anything like that. It follows the speed or the momentum of price. As a rule, the momentum changes direction before price."[8]

Divergences within the stochastic oscillator are believed to predict reversal and precede it, giving traders an excellent timing mechanism to make trades before price movement. Two separate moving averages are used in stochastics: these are called %k and %d.

To calculate each, the high and low prices refer to the "look-back" period, usually 14 sessions:

%K: (current close – lowest low) ÷ (highest high – lowest low) x 100

%D: 3-day simple moving average of %k

Because the stochastic oscillator is intended as a measure of the two averages (14-day close/low divided by high/low and 3-day simple moving average), the oscillator appears as two closely aligned waves on a price chart. When a price closes above 50, it means the close was in the upper half of the range; if it closes below 50, the close was in the lower half. Very low readings (under 20, for example) indicate that price is near a trend low, and very high readings (above 80 for most interpretations) indicate that price is at or near its trend high. In either case, reversal would be most likely in coming sessions.

Tip: Find a momentum oscillator that works best for you. But don't rely only on confirmation. Contradiction—when the oscillator disagrees with the price signal—is an equally strong indicator of coming change.

Stochastic oscillators may come in one of three varieties. Fast stochastic oscillators are based on the original formula. Slow stochastic oscillators smooth out the otherwise choppy three-day simple moving average, which under the fast variety might appear as a highly volatile attribute.

A full stochastic oscillator (which is used in the following chart) enables traders to adjust the look-back period to something other than 14 sessions. The full version is shown on the chart of Exxon Mobil in figure 12.8. The lighter of the two lines lags behind the darker line and represents the three-day simple moving average. It tracks price closely throughout the period charted, demonstrating that no rapid trending movements are occurring; in such instances, the two stochastic lines could widen, but that would be short-term as the 3-day average is catching up with the longer-term 14-day line.

Note that when price moved above resistance, the stochastic oscillator moved above its 80 mark, signaling a likely overbought

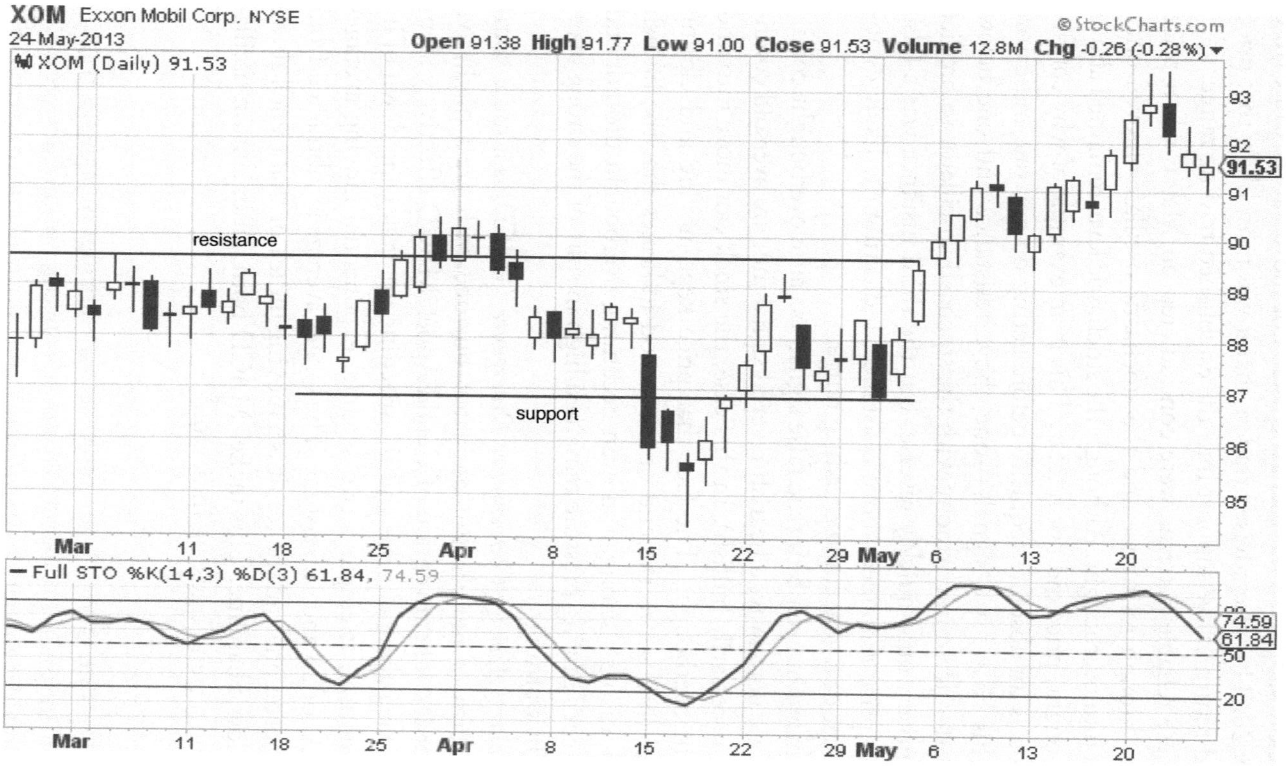

Figure 12.8 Exxon Mobil, Stochastic Oscillator
Source: Chart courtesy of StockCharts.com

condition, and warning traders that this was not likely to be a lasting breakout. The same pattern was seen on the low side when price fell below support. The stochastic oscillator moved at the same time below 20. This also advised traders that it was most likely that price would return above support into the established trading range.

These reliable patterns were contradicted on May 6 when price broke through resistance and the stochastic oscillator moved once again into overbought. Although the stochastic level remained above overbought, price did not retreat. This is an example of why confirmation is an absolute requisite before acting on any one indicator. On this chart, two signals—one above resistance and one below support—acted in harmony with the stochastic oscillator to predict that price would quickly reverse. But the third instance was a false signal; price remained above resistance, indicating the likely establishment of a new, higher trading range.

It may not even be a true false signal, because overbought conditions are not always a bearish reversal signal (nor are oversold conditions necessarily bullish). In this case, the stock can move into and remain overbought. The fact that price kept moving and closing higher through to the end of the period indicates that the bullish trend was very strong, and even becoming overbought did not end the trend. Buying pressure remained high in spite of the strong overbought warning.

Moving Average Convergence Divergence (MACD) is a more complex form of momentum oscillator, but one with great predictive properties. The invention of Gerald Appel, MACD effectively identifies the pacing of price and the speeding up or slowing down of its momentum. It employs two separate exponential moving averages (EMA) of 26 and 12 sessions. The longer average is subtracted from the shorter one and the result is a moving indicator that trends above or below a baseline. This makes MACD both a momentum oscillator and a trend tracking indicator.[9]

> **Key point:** MACD combines several moving averages to track momentum in the form of crossover, convergence, and divergence. It provides a wealth of information and anticipates price reversal in many instances.

Meaning is derived when the moving averages converge, diverge, or cross the centerline, which is fixed. The formula for MACD is based on its three separate averages:

MACD line: 12-day EMA – 26-day EMA
Signal line: 9-day EMA of the MACD line
MACD histogram: MACD line – signal line

The histogram provides a positive indication when the MACD line is above its signal line, and it is negative when the MACD line is below its signal line. As the moving averages converge or diverge, a signal is produced. Divergence is recognized as the averages move away from one another, and convergence is seen as the two averages approach one another.

The centerline or zero line is fixed. Positive MACD is found when the faster 12-session EMA is higher than the slower 26-session EMA. The more divergence is seen between these two, the more positive the indication gained from the signal. Upside momentum is increasing when this occurs.

Bearish signals include the 12-session EMA trending below the 26-session EMA. The farther the divergence between the two, the greater the bearish signal. At this point, downside momentum is growing.

A signal line crossover is the best-known MACD signal. The signal line is limited to nine consecutive sessions. Bullish crossover is identified when MACD turns upward and crosses above the signal line; bearish crossover is the opposite, when MACD turns down and moves below the signal line. The more time spent in either territory (bullish or bearish), the stronger the MACD signal is likely to be.

A related important crossover involves the fixed centerline. A bullish centerline crossover occurs when MACD moves above the line, and bearish signals include crossover below the signal line.

MACD divergence occurs whenever the moving averages move away from price levels. A bullish divergence is experienced when the stock price reaches a lower low price than in the previous session, but MACD forms a higher low level. The price action reveals a downtrend, but the diverging MACD reveals a slowing down of that downside trend. Bearish divergence consists of a stock reaching a higher high price, while MACD reaches a lower high. The price action reveals continuing uptrend action, but the MACD trend shows that the momentum of that bearish move is slowing.

> **Tip:** Divergence usually tells you that the price trend's momentum is slowing down, often the first sign of coming reversal.

Divergence in both forms indicates that momentum is slowing and may anticipate a reversal and price movement in the opposite direction. More than other momentum oscillators, MACD and its separate EMA trends are a strong confirming signal. Although divergence is subtle in relation to outright price reversals, it also provides an exceptionally reliable early forecast of coming changes in price trends. Although divergence is common as part of a trend, in which retracement occurs often without changing the longer-term trend, MACD may give insight concerning not the price direction, but its speed and duration. Divergence that does not disappear but continue may signal the end of the existing trend.

The chart of Chevron reveals MACD and its components; this combined price and MACD chart is shown in figure 12.9.

Note that the candlestick reversals occur at about the same point of MACD crossover, convergence, or directional trend. The first of these is the bearish crossover (12-session EMA crossover below 26-session EMA) at the same point as the bearish engulfing pattern.

A brief uptrend concludes as convergence occurs but does not lead to crossover. At the height of this uptrend, a bearish harami was also spotted. This also confirms the downward reversal.

The third signal is contradictory and is a good example of how MACD may give out a false reading. The downtrend concluded with a bullish harami; however, the MACD indications were all bearish. This would be confusing in the moment, but the chart demonstrates that in spite of MACD's signals, price did reverse strongly to the upside. As the EMA lines fell below the signal line, you would recognize this as quite bearish, but price did not cooperate.

The final signal consisted of trend lines moving strongly upward along with MACD signals of a positive trend. However, at the break point between the two upward trend lines, note that the MACD lines converge, indicating that the uptrend momentum was falling.

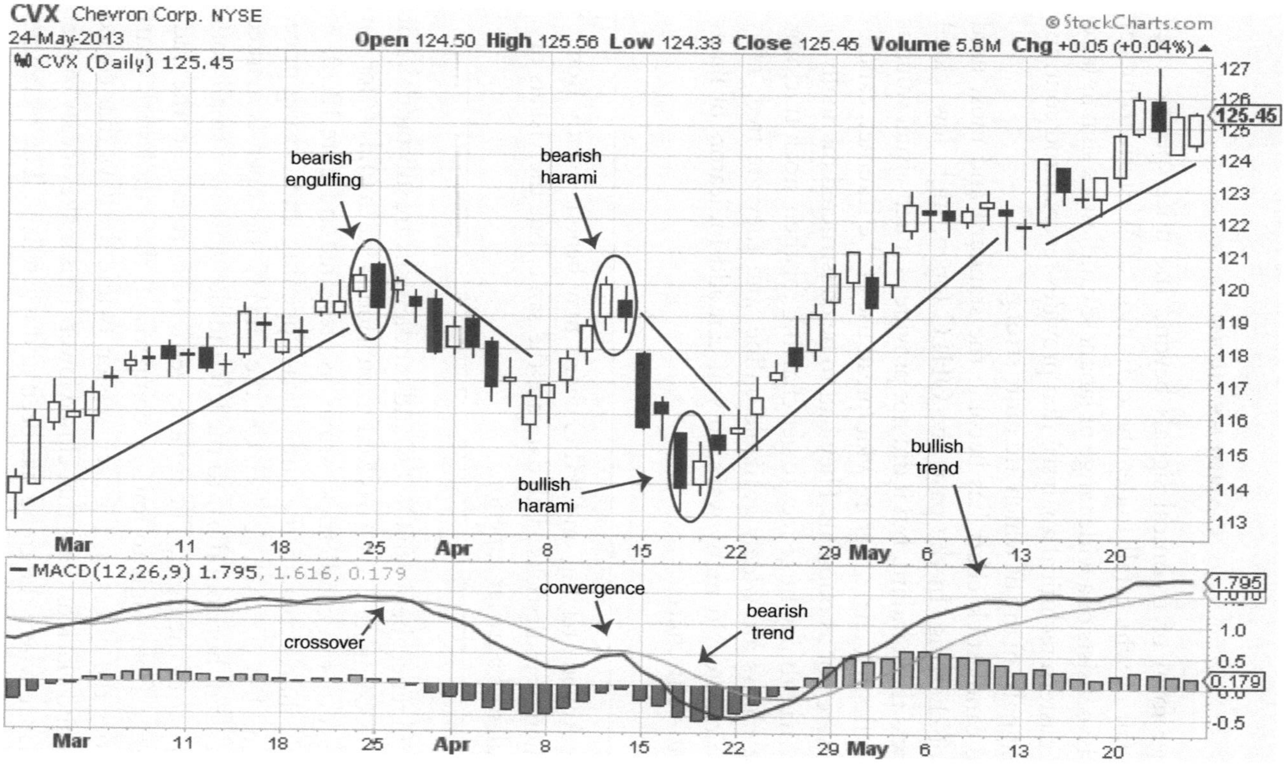

Figure 12.9 Chevron, Moving Average Convergence Divergence
Source: Chart courtesy of StockCharts.com

This was not a bearish signal, but rather a slowing down in the prevailing bullish price direction.

CANDLESTICKS

Another form of technical signals beyond the price-specific ones above (all of which fall under the definition of Western technical analysis) consists of a wide range of Japanese candlestick signals (also called Eastern technical analysis) that provides reversal and continuation signals, many of which are quite strong.

The candlestick chart has become the standard and most used form of charting. In the past, the OHLC (open, high, low, close) chart was the standard, but it did not provide as much visual value as the candlestick chart. Before widespread use of the Internet, candlesticks were not as popular. They had to be constructed manually, or traders had to pay huge fees to get charting for their stock tracking. Today, charts are widely available online, and they are free. They can be tailored in many ways, so you can add indicators of many types to your charts.

Tip: Anyone relying on OHLC charts may benefit greatly from studying candlesticks. The data are the same, but the visuals of candlesticks are far superior.

The differences between OHLC and candlesticks are substantial. For example, compare two charts for Exxon Mobil, one in OHLC format and the other in candlestick format, as presented in figure 12.10.

Both charts contain the same data—opening and closing price, high and low for each session. On the OHLC, the open is marked with a small horizontal bar to the left and the close by a small horizontal bar to the right. The middle, vertical bar represents the trading range from high to low. In comparison, the candlestick chart identifies the direction for each session (white for days that closed higher and black for days that closed lower. This makes it easy to spot at a glance where the trend has been and where it is. The rectangle, called the real body, represents the range between open and close (on a white session, the open is at the bottom and the close at the top; on a black session, it is the other way around). The extensions above

OHLC chart

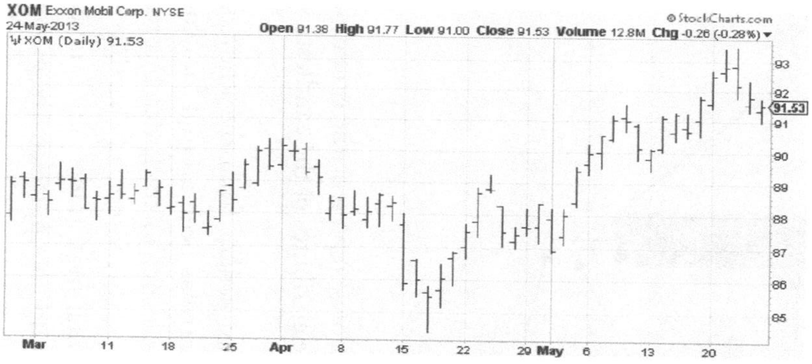

Candlestick chart

Figure 12.10 Exxon Mobil OHLC and candlestick charts
Source: Charts courtesy of StockCharts.com

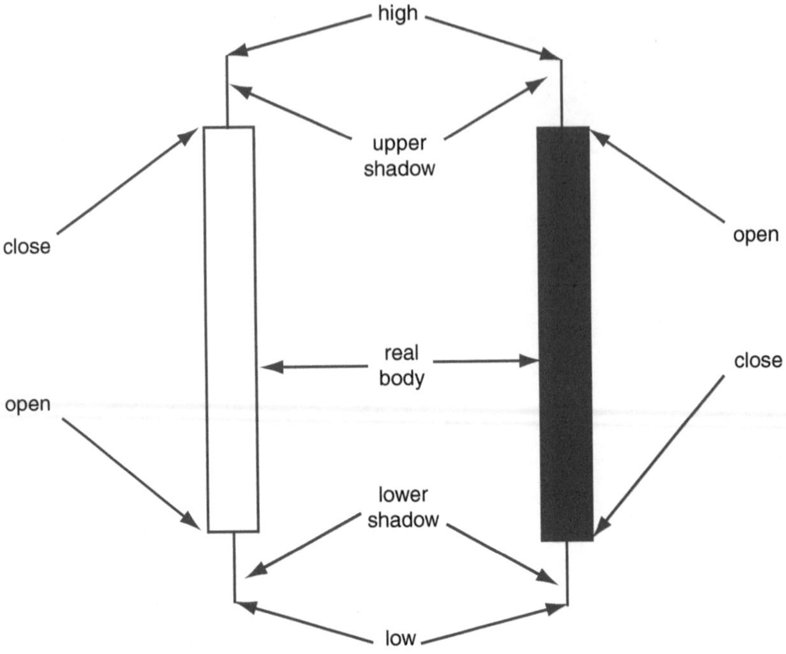

Figure 12.11 The candlestick and its parts
Source: Created by the author

and below, called upper and lower shadows, represent the extent of the full trading range for the day. These attributes of the candlestick chart are summarized in figure 12.11.

All investor in the energy market can benefit from the study of candlestick charts. The candlestick chart has been around for centuries, and was used in the sixteenth century to track rice futures in Japan. However, in the United States, this format of charting is quite new. It is reasonable to estimate that most traders use candlestick charts for their analysis, but most do not appreciate the power of candlestick reversal and continuation patterns. Charts displayed earlier in this chapter highlighted some of the more important patterns, but these are only a few. Following are some of the most significant candlestick patterns shown in single-session, double-session, and three-session configurations.

Single-Session Candlesticks

Long candlesticks are long relative to other sessions. In the illustration, two typical sessions are shown, followed by a white (bull) and a black (bear) long candlestick. The long candlestick may vary based on scale. For example, if price is scaled in 5-point increments, a 5-point move might not appear very long, but if the chart is scaled in 1-point increments, a 5-point move will form a long candlestick session. With this in mind, "long" is a matter of the relationship between typical sessions and an unusually extended one, as the example in figure 12.12 shows.

Four types of small-body candlesticks are highlighted in figure 12.13. These are the spinning top, doji, hanging man, and hammer.

> **Tip:** Candlestick indicators often have colorful names, such as doji (which means mistake). The names tell you how to interpret the signals.

The spinning top is a session with a small real body, likely to be more square than rectangular; and large upper *and* lower shadows. The shadows on both sides should be at least as long as the real body, and the longer the shadows, the stronger the indicator. The real body may be white or black, and its significance depends on where it is located. If it appears at the top of an uptrend, it is likely to serve as a bear reversal; when seen at the bottom of a downtrend, it may be a bullish reversal signal.

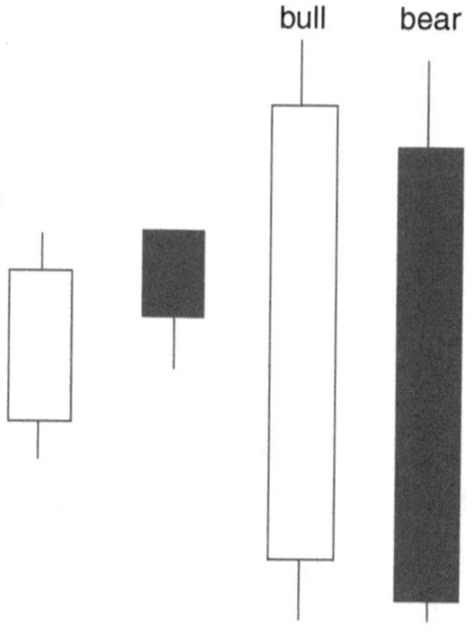

Figure 12.12 Long candlesticks
Source: Prepared by author

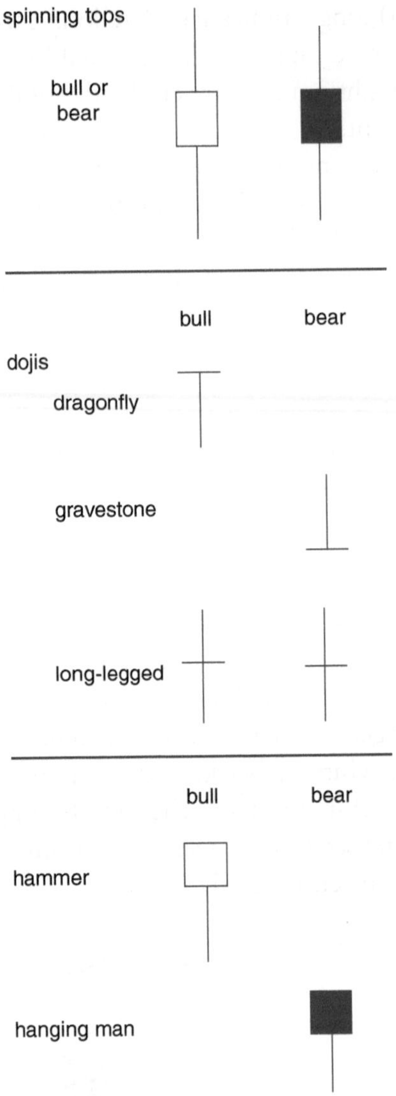

Figure 12.13 Small-body candlesticks
Source: Prepared by author

Doji sessions come in many varieties. Among these are three specialized ones: the bullish dragonfly (with the horizontal line on the top and a lower shadow only), the bearish gravestone (with the horizontal line at the bottom and an upper shadow only), or the long-legged doji, with the horizontal line in the middle and exceptionally long upper and lower shadows. The long-legged doji may be bullish or bearish, depending on where it shows up within the current trend.

DOUBLE-SESSION CANDLESTICKS

There are many types of reversal and continuation two-session candlesticks. Here, only five are highlighted.

The engulfing pattern consists of the second session with a real body extended both higher and lower than the real body of the first session. This is one of the strongest candlestick reversal signals, leading to actual trend reversal about 63% of the time (bullish) or 79% of the time (bearish).[10]

The bullish engulfing begins with a black session and is followed by a larger white session that "engulfs" the first session. The bearish is the opposite, with a white session engulfed in the second session by a larger black candlestick.

The harami is the opposite configuration of the engulfing pattern. It starts with a large session and is followed by a session of the opposite color that is smaller on both the top and the bottom. The bullish harami starts with a black session and is followed by a white one; the bearish harami begins with a white session and is followed by a black session.

The harami is reliable only about 50% of the time, so by itself this is not an especially valuable indicator. However, when it confirms separate reversal signals (other candlesticks, gapping price action, double tops or bottoms, volume spikes, or changes in momentum oscillators), it adds to the likelihood that reversal will occur.

> **Tip:** In Japanese, *harami* means "pregnant," a reference to the appearance of the two-session indicator.

Meeting lines are characterized by the closing prices at the same level. In the bullish version, a black session closes down, and the following day it opens at a downside gap but returns to close at the same price as the first session. In the bearish version, the first day is a white session. The second session gaps to open higher, but then it declines to close at the same price as the previous session.

Piercing lines extend the meeting lines principle. The gap between the two sessions remains an important characteristic. However, the second session closed higher (bullish) or lower (bearish) than the closing price of the first session.

The four highlighted bull and bear versions of these double-session candlestick patterns are summarized in figure 12.14.

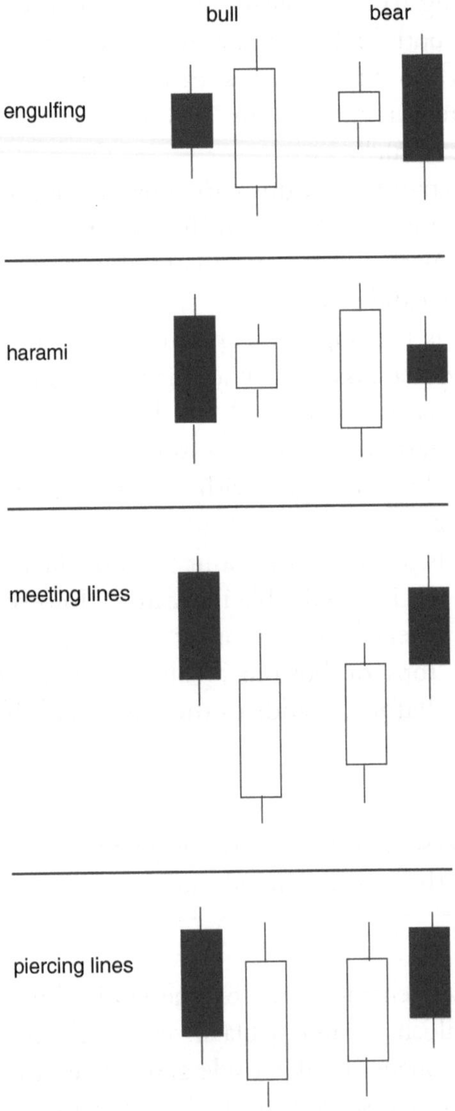

Figure 12.14 Double-session candlesticks
Source: Prepared by author

TRIPLE-SESSION CANDLESTICKS

Like double-session indicators, triple-session indicators come in many varieties of reversal or confirmation candlesticks. Only three are highlighted here.

Three white soldiers and three black crows are signals that may indicate reversal or act as confirmation, depending on where they appear. If the three white soldiers pattern appears during a downtrend, it probably indicates a reversal; if it is found during an uptrend, it confirms the trend. The pattern of three black crows is the opposite and may also serve to indicate reversal (showing up in an uptrend) or to confirm (showing up in a downtrend).

The three white soldiers pattern consists of three consecutive white sessions. Each one opens with a higher low and closes with a higher high. The three black crows pattern is made up of three consecutive black sessions. Each of these sessions opens with a lower high and closes with a lower low.

Key point: Candlesticks are not exclusively reversal or confirmation signals. Their function depends on where an indicator shows up in the current trend.

The abandoned baby is a three-session indicator. The bullish version starts with a black session and then gaps downward and is followed by a doji in the second session. Then price gaps upward, and the third session concludes with a white candlestick. In the bearish version, the first session is white and is followed by an upward gap. The second session is a doji followed by a downward gap. The third session is a black candlestick.

The squeeze alert is an unusual pattern because only the first session's color matters. The second and third session may be white or black. Another way in which these patterns are unusual is that the black squeeze is bullish, and the white squeeze is bearish. In the bullish version, each session opens and closes within the size of the previous session's real body, that is, it is squeezed smaller. The first session is white, and preferably sessions two and three will also be white, but this is not essential. The black squeeze alert has the opposite configuration, consisting of three sessions, each one smaller than

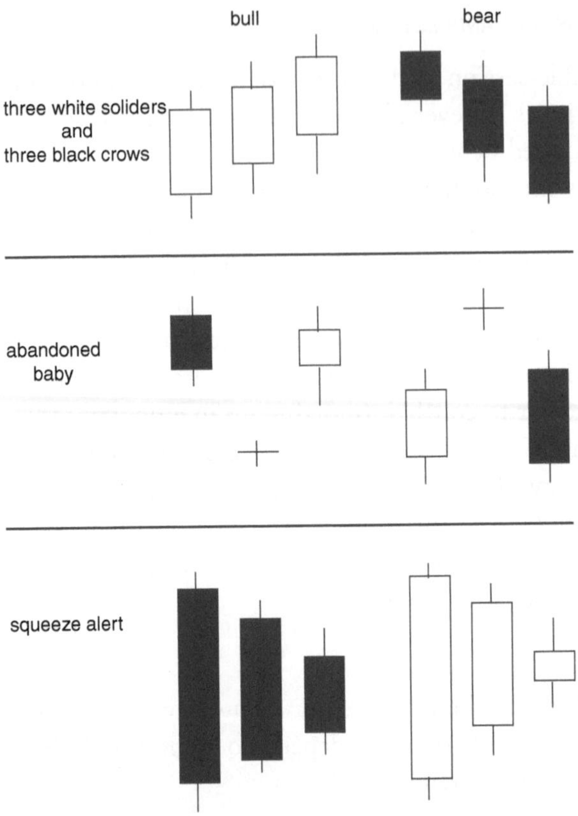

Figure 12.15 Triple-session candlesticks
Source: Prepared by author

the previous one. The ideal bear squeeze alert has three white sessions, but sessions two and three may be either white or black. All three of these triple-session candlestick formations are summarized in figure 12.15.

Testing and Analyzing Price

Technical analysis often is performed as a solitary exercise. You can locate free charting services online and track any listed energy company and see how its chart patterns develop. This is not limited to a study of price patterns, but you may include volume, a broad range of technical indicators, and momentum oscillators in your research. The website used for charts in this book, StockCharts.com, is free, and you can set up charts in a variety of ways—including duration, type of indicators, and placement on the chart (above price, below price, or behind price, for example).

Tip: Finding free charting services is easy online, and you can tailor charts to include as much or as little data as you need.

You will become skilled at interpreting technical charts with study and practice. Because there are so many different technical signals, it makes sense to limit usage to a very short list of indicators that work for you. For example, you may use candlestick charts and look for reversal in price patterns and then confirm these with other price signals (Western and Eastern), volume spikes or indicators, and one or two momentum oscillators.

Given the higher than average volatility of energy stocks, technical analysis is a smart way to look for bargains or to spot overpriced stocks. When combined with fundamental analysis, this rounds out your ability to pick stocks and time you entry and exit. Market analysts may believe strongly in the value of technical analysis to spot emerging trends and often to anticipate a reversal before it occurs. The reward in technical analysis is found when indicators are confirmed and used to time entry and exit to maximize profits.

CHAPTER 13

How Energy Trades Are Taxed

> To tax and to please, no more than to love and to be wise, is not given to men.
>
> Edmund Burke, *On American Taxation*, 1775

The US income tax law is a complex and often illogical series of rules, regulations, and court case decisions. Few individuals are able to decipher this complexity; most traders thus limit their knowledge to those rules affecting them directly and then rely on a tax accountant or attorney to help with the rest.

This chapter very briefly summarizes the major rules affecting energy investors. Every energy investor needs to know the following tax-based definitions:

Ordinary income is also known as *earned income*. It includes all wages or salaries, income from a business, interest, dividends, and commissions but excludes capital gains. A distinction is made between "ordinary" and "earned," however, that goes beyond this definition. If you receive an advance from your employer, for example, it is "unearned" because it is not taxed until you perform the work related to the income. However, when it becomes earned it will be taxed as ordinary income; this means you will pay taxes on this amount based on the effective tax rate applied to your ordinary income.

Key point: Ordinary income is taxed at the effective tax rate; capital gains are taxed depending on your income level.

Capital gains are taxed at different rates. A short-term gain (on assets held less than one year) is taxed like ordinary income, but a long-term capital gain is taxed at a lower rate. Congress has amended this rate many times over the past several years, so the rate in effect today depends on the latest amendment to the tax code and on current ordinary rates.

Investment income includes all profits and losses from investment activity: capital gains, dividends, and interest. Tax treatment for most forms of investment income is the same as for other types of ordinary income, except for long-term capital gains. However, capital losses cannot be deducted without limit. The maximum annual net loss deduction is limited to $3,000 per year for a married couple; any excess then has to be carried forward to future years and applied against future capital gains.[1]

> **Tip:** The new net investment income tax affects only those whose income is above $250,000 (married) or $200,000 (single).

As of 2012, a new tax was passed by Congress. All net investment income is taxed at an additional 3.8% above other income tax rates. This does not affect most people; only those with income above $250,000 (married couples) or $200,000 (single people) will be assessed this additional tax.[2]

Passive income is any income derived from activities in which there is no "material participation." That is, if you are a partner in a business but do not manage it directly, your income from that business is probably passive. If you own shares in a passive investment and it loses money, a passive loss is probably not deductible but has to be carried over and applied to future passive gains. In the past, passive losses could be deducted, and this led to a lot of abuse. High-rate taxpayers invested in limited partnerships, for example, and wrote off their losses against income. These were classified by how much of a multiplier could be deducted. For example, a "four to one" loss meant that a $100,000 investment generated a $400,000 net loss. When this was deducted from other income, the loss reduced tax liabilities beyond the amount invested. This is why passive losses can no longer be deducted; the abusive tax shelters of past years created ways for people to avoid taxes even when their income was quite high. Examples of passive income today include rent income, royalties, and limited partnership profit or loss. A distinction is also made between passive and portfolio income. To a degree, portfolio income is the same as investment income and is taxed at ordinary rates (except for long-term capital gains). However, losses of portfolio income are deductible (subject to the capital gains limitation of $3,000 per year), whereas most other passive losses are not deductible.[3]

A related rule was also passed once limitations on passive loss went into effect. This is the "at-risk rule." As a general rule, you cannot deduct any amounts above the amount you have at risk. This means the amount invested, loaned, or used in the activity. For example, energy investments may generate losses, but you can never deduct more than you invest; the maximum loss is equal to 100% of your at-risk dollar value. If you buy 100 shares on margin, the entire amount is at risk because your brokerage firm has the right to recover the 50% it lends to you even if the entire investment becomes worthless.

Your *effective tax rate* is the rate of taxes you pay. It is not based on your total income, but on your net taxable income after adjustments to gross income, deductions, and exemptions. To find your effective tax rate, divide your total tax liability by taxable income. In 2012, on the federal Form 1040, taxable income was shown on page 2, line 43, and tax liability was on line 46. However, your overall effective tax rate should also allow for state tax liabilities. To find this, one calculation involves adding together your federal and state tax liabilities and dividing the total by the federal taxable income.

Key point: Your tax liability is calculated based on your marginal tax rate, which increases as your taxable income rises.

Effective tax rate is more important than the better-known *marginal tax rate*, which is the rate assessed by taxing authorities against tiers of income. Thus, the percentage of the effective tax rate rises as the marginal rate rises. A sample of the marginal rates for 2012 is shown in table 13.1.

Table 13.1 Marginal tax rates, 2012

Taxable income				
Over:	**but not over:**	**Tax**	**+**	**over**
$ 0	$ 17,400	$ 0	10%	$ 0
17,400	70,700	1,740	15%	17,400
70,700	142,700	9,734	25%	70,700
142,700	217,450	27,735	28%	142,700
217,450	388,350	48,665	33%	217,450
388,350		105,062	35%	388,350

Source: Internal Revenue Service (IRS), at www.irs.gov/pub/irs-pdf/i1040tt.pdf.

STATE INCOME TAXES

To find the effective tax rates for each state, it is necessary to understand how income is taxed in your state of residence. A summary of state income tax rules is provided in Table 13.2.

Finding your effective tax rate by state is complicated further by the fact that many states' rates include income brackets, exemptions, and other adjustments of the tax liability. For example, six states allow deduction of federal income taxes in arriving at taxable income (Alabama, Iowa, Louisiana, Missouri, Montana, and Oregon). In many states, income brackets exclude lower income levels from tax, so there is no simple way to peg the effective rate without performing the calculation.

> **Tip:** Your state income tax liability depends on rules in effect and on the rates, income brackets, and exemptions each state allows.

One sound reason for knowing the combined federal and state rate is to calculate the level of return you need to break even. This is your after-tax rate of return. For example, if you earned 4.5% in your portfolio last year, but your combined federal and state effective tax rate is 45.3%,[4] then your after-tax return is approximately 2.5%:

4.5% x (100% – 45.3%) = 2.4615%

This is considerably lower than the pretax yield on your portfolio. Making matters even more serious, if you assume that inflation is

Table 13.2 State income tax rates, 2013

No state income tax	
Alaska	Florida
Nevada	New Hampshire *
South Dakota	Tennessee †
Texas	Washington
Wyoming	
Flat rates:	
Massachusetts, 5.3%	Illinois, 5%
Utah, 5%	Colorado, 4.63%
Michigan, 4.25%	Indiana, 3.4%
Pennsylvania, 3.07%	
By highest maximum rates	
California, 13.3%	Hawaii, 11%
Oregon, 9.9%	Iowa, 8.98%
New Jersey, 8.97%	DC, Vermont, 8.95%
New York, 8.82%	Maine, 8%
Minnesota, 7.85%	North Carolina, Wisconsin, 7.75%
Idaho, 7.4%	Arkansas, South Carolina, 7%
Montana, 6.9%	Nebraska, 6.84%
Delaware, 6.75%	Connecticut, 6.7%
West Virginia, 6.5%	Georgia, Kentucky, Louisiana, Missouri 6%
Rhode Island, 5.99%	Ohio, 5.925%
Maryland, 5.8%	Virginia, 5.75%
Oklahoma, 5.25%	Alabama, Mississippi, 5%
Kansas, New Mexico, 4.9%	Arizona, 4.54%
North Dakota, 3.99%	

*5% tax on dividends and interest income only

† 6% tax on certain dividends and interest income only

Source: Federation of Tax Administrators, January 2013, at www.taxadmin.org.

averaging 2% per year, the net after-tax and after-inflation net is a *loss*. To calculate, divide the inflation rate by the after-tax rate. Assuming a 3% rate of inflation ("I" is the inflation rate, "E" is the effective tax rate, and "B" is the breakeven rate):

$$I \div (100 - E) = B$$

Based on 3% inflation and an effective tax rate of 45.3%:

$$3\% \div (100 - 45.3\%) = 5.5\%$$

You would need to earn 5.5% just to break even under these conditions. Even though the initial rate of 4.5% might seem attractive, when you calculate the effects of taxes and inflation, you discover that you need an additional 1% just to keep your after-tax, after-inflation value. If you earn less than 5.5%, you are not breaking even.

This is the dilemma for energy investors. To earn 5.5% you have to assume greater risks or find some way to increase current income. That is why the combination of higher than average dividend yield plus conservative options strategies solves the risk/return problems of stock ownership. Unless a stock rises significantly each and every year, you cannot break even consistently. However, with high dividend yield and options strategies, you can beat the required breakeven level without incurring higher risks.

More Tax Rules: The Wash Sale Danger

If you want to time your trades for year-end planning, be aware of the *wash sale* rule. For example, if you hold stock that has lost value, you might want to sell shares before December 31 to take the loss this year and then repurchase shares in the new year. This is a great plan to create a reduction in current-year taxes, but if you repurchase the shares within 30 days, the loss in the current year is negated as a wash sale. That is, if you move back into the position within 30 days, you cannot claim the deduction. All you get from a wash sale is the cost of trades and the risk that you will lose money on the trade.

Key point: Any effort to close and then reopen a position within 30 days is probably a wash sale—including replacement of stock with options to offset the same risks.

Some traders thought they could outwit the rule by selling shares and replacing them with options. The idea here is that any movement in the stock is offset by the option, but this does not work either. The wash sale recognizes options as "substantially identical stock or securities" and the wash sale rule still applies.[5]

SHORT TERM OR LONG TERM

Capital gains are short-term if the position was open for one year or less; these are taxed at the same rate as ordinary income. Long-term capital gains apply to investments open for more than one year. As of 2012, a complex series of long-term rates applies:

0% rate: applies to anyone whose income is taxed in the 10% or 15% bracket; also applies to home sales for those who used homes as primary residences at least two of the last five years and is limited to gains of $500,000 (married couples) or $250,000 (single people).

15% rate: applies to anyone whose income is taxed in the 25% bracket.

20% rate: applies to anyone with taxable income above $400,000.

25% rate: applies to property owners taxed at the 25% rate (applies to real estate profits).

28% rate: applies to anyone in the 28% tax bracket or higher (tax applies to collectibles and small business stock profits).

3.9% Medicare Surtax: applies to capital gains for anyone whose gross income is higher than $200,000.[6]

Capital gains, both short-term and long-term, are reported on the federal Schedule D. Page 1 of Schedule D is shown in figure 13.1.

This form, for tax year 2012, has two primary sections, one for short-term and one for long-term capital gains. These are transferred to the form from Form 8949, which is simply a listing of gains or losses in each category.

Page 2 of Schedule D calculates the net reportable income by combining short-term and long-term gains and—depending on whether each is a profit or a loss—the amount to be carried over to the main Form 1040 and reported with other income. Page 2 is shown in figure 13.2.

The summarized totals of short-term and long-term gains are transferred to Schedule D from Form 8949, which is also a two-page form. The first page of this form is shown in figure 13.3.

This page is used solely for the listing of all short-term gains and losses on assets owned for one year or less. On page 2, the long-term gains or losses are summarized, as shown in figure 13.4.

SCHEDULE D (Form 1040)

Department of the Treasury
Internal Revenue Service (99)

Capital Gains and Losses

▸ Attach to Form 1040 or Form 1040NR.
▸ Information about Schedule D and its separate instructions is at *www.irs.gov/form1040*.
▸ Use Form 8949 to list your transactions for lines 1, 2, 3, 8, 9, and 10.

OMB No. 1545-0074

2012

Attachment Sequence No. **12**

Name(s) shown on return | **Your social security number**

Part I **Short-Term Capital Gains and Losses—Assets Held One Year or Less**

Complete Form 8949 before completing line 1, 2, or 3. This form may be easier to complete if you round off cents to whole dollars.	(d) Proceeds (sales price) from Form(s) 8949, Part I, line 2, column (d)	(e) Cost or other basis from Form(s) 8949, Part I, line 2, column (e)	(g) Adjustments to gain or loss from Form(s) 8949, Part I, line 2, column (g)	(h) Gain or (loss) Subtract column (e) from column (d) and combine the result with column (g)
1 Short-term totals from all Forms 8949 with **box A** checked in **Part I**				
2 Short-term totals from all Forms 8949 with **box B** checked in **Part I**				
3 Short-term totals from all Forms 8949 with **box C** checked in **Part I**				

4 Short-term gain from Form 6252 and short-term gain or (loss) from Forms 4684, 6781, and 8824 .	4	
5 Net short-term gain or (loss) from partnerships, S corporations, estates, and trusts from Schedule(s) K-1	5	
6 Short-term capital loss carryover. Enter the amount, if any, from line 8 of your **Capital Loss Carryover Worksheet** in the instructions	6	( )
7 **Net short-term capital gain or (loss).** Combine lines 1 through 6 in column (h). If you have any long-term capital gains or losses, go to Part II below. Otherwise, go to Part III on the back	7	

Part II **Long-Term Capital Gains and Losses—Assets Held More Than One Year**

Complete Form 8949 before completing line 8, 9, or 10. This form may be easier to complete if you round off cents to whole dollars.	(d) Proceeds (sales price) from Form(s) 8949, Part II, line 4, column (d)	(e) Cost or other basis from Form(s) 8949, Part II, line 4, column (e)	(g) Adjustments to gain or loss from Form(s) 8949, Part II, line 4, column (g)	(h) Gain or (loss) Subtract column (e) from column (d) and combine the result with column (g)
8 Long-term totals from all Forms 8949 with **box A** checked in **Part II**				
9 Long-term totals from all Forms 8949 with **box B** checked in **Part II**				
10 Long-term totals from all Forms 8949 with **box C** checked in **Part II**				

11 Gain from Form 4797, Part I; long-term gain from Forms 2439 and 6252; and long-term gain or (loss) from Forms 4684, 6781, and 8824	11	
12 Net long-term gain or (loss) from partnerships, S corporations, estates, and trusts from Schedule(s) K-1	12	
13 Capital gain distributions. See the instructions	13	
14 Long-term capital loss carryover. Enter the amount, if any, from line 13 of your **Capital Loss Carryover Worksheet** in the instructions	14	( )
15 **Net long-term capital gain or (loss).** Combine lines 8 through 14 in column (h). Then go to Part III on the back	15	

For Paperwork Reduction Act Notice, see your tax return instructions. Cat. No. 11338H **Schedule D (Form 1040) 2012**

Figure 13.1 Federal Tax Schedule D, Capital Gains and Losses (page 1)
Source: www.irs.gov

Once the list of gains and losses is moved to Schedule D, the taxable (or deductible) dollar value is computed. This includes carrying over any losses from previous years. Although net gains are taxable each year, losses are limited to $3,000. Any losses above $3,000 have to be carried over and applied to future years. This does not mean that it will take a full 10 years to absorb a $30,000 loss, however. The carryover is applied, which means that if you have a year with a substantial gain, the carryover value is deducted from that gain. The entire loss could be absorbed in only a few years of capital gains. For example, if your carryover loss this year is $14,500 and you report

Schedule D (Form 1040) 2012 Page **2**

Part III Summary

16 Combine lines 7 and 15 and enter the result **16**

- If line 16 is a **gain**, enter the amount from line 16 on Form 1040, line 13, or Form 1040NR, line 14. Then go to line 17 below.
- If line 16 is a **loss**, skip lines 17 through 20 below. Then go to line 21. Also be sure to complete line 22.
- If line 16 is **zero**, skip lines 17 through 21 below and enter -0- on Form 1040, line 13, or Form 1040NR, line 14. Then go to line 22.

17 Are lines 15 and 16 **both** gains?
☐ **Yes.** Go to line 18.
☐ **No.** Skip lines 18 through 21, and go to line 22.

18 Enter the amount, if any, from line 7 of the **28% Rate Gain Worksheet** in the instructions . . ▶ **18**

19 Enter the amount, if any, from line 18 of the **Unrecaptured Section 1250 Gain Worksheet** in the instructions . . . . ▶ **19**

20 Are lines 18 and 19 **both** zero or blank?
☐ **Yes.** Complete the **Qualified Dividends and Capital Gain Tax Worksheet** in the instructions for Form 1040, line 44 (or in the instructions for Form 1040NR, line 42). **Do not** complete lines 21 and 22 below.

☐ **No.** Complete the **Schedule D Tax Worksheet** in the instructions. **Do not** complete lines 21 and 22 below.

21 If line 16 is a loss, enter here and on Form 1040, line 13, or Form 1040NR, line 14, the **smaller** of:
- The loss on line 16 or
- ($3,000), or if married filing separately, ($1,500) } . . . . **21** ( )

Note. When figuring which amount is smaller, treat both amounts as positive numbers.

22 Do you have qualified dividends on Form 1040, line 9b, or Form 1040NR, line 10b?

☐ **Yes.** Complete the **Qualified Dividends and Capital Gain Tax Worksheet** in the instructions for Form 1040, line 44 (or in the instructions for Form 1040NR, line 42).

☐ **No.** Complete the rest of Form 1040 or Form 1040NR.

Schedule D (Form 1040) 2012

Figure 13.2 Federal Tax Schedule D, Capital Gains and Losses (page 2)

gains of $11,000, none of this gain will be taxable due to the carryover. The remaining loss of $2,500 ($14,500 – $11,000) is carried over and applied next year.

> **Tip:** A carryover loss can be used to take gains and avoid taxes on them, because the gains are offset by the carryover. This means that even a large carryover loss does not necessarily take several years to get absorbed.

Also listed on Schedule D are *capital gain distributions*. These are gains reported by mutual funds or real estate investment trusts,

Form **8949**

Department of the Treasury
Internal Revenue Service

Sales and Other Dispositions of Capital Assets

▶ Information about Form 8949 and its separate instructions is at *www.irs.gov/form8949*.
▶ File with your Schedule D to list your transactions for lines 1, 2, 3, 8, 9, and 10 of Schedule D.

OMB No. 1545-0074
2012
Attachment Sequence No. **12A**

Name(s) shown on return | **Social security number or taxpayer identification number**

Most brokers issue their own substitute statement instead of using Form 1099-B. They also may provide basis information (usually your cost) to you on the statement even if it is not reported to the IRS. Before you check Box A, B, or C below, determine whether you received any statement(s) and, if so, the transactions for which basis was reported to the IRS. Brokers are required to report basis to the IRS for most stock you bought in 2011 or later.

Part I **Short-Term.** Transactions involving capital assets you held one year or less are short term. For long-term transactions, see page 2.

You *must* check Box A, B, *or* C below. Check only one box. If more than one box applies for your short-term transactions, complete a separate Form 8949, page 1, for each applicable box. If you have more short-term transactions than will fit on this page for one or more of the boxes, complete as many forms with the same box checked as you need.

☐ **(A)** Short-term transactions reported on Form(s) 1099-B showing basis **was** reported to the IRS
☐ **(B)** Short-term transactions reported on Form(s) 1099-B showing basis was **not** reported to the IRS
☐ **(C)** Short-term transactions not reported to you on Form 1099-B

1 (a) Description of property (Example: 100 sh. XYZ Co.)	(b) Date acquired (Mo., day, yr.)	(c) Date sold or disposed (Mo., day, yr.)	(d) Proceeds (sales price) (see instructions)	(e) Cost or other basis. See the **Note** below and see *Column (e)* in the separate instructions	**Adjustment, if any, to gain or loss.** If you enter an amount in column (g), enter a code in column (f). **See the separate instructions.** (f) Code(s) from instructions	(g) Amount of adjustment	(h) **Gain or (loss).** Subtract column (e) from column (d) and combine the result with column (g)
2 Totals. Add the amounts in columns (d), (e), (g), and (h) (subtract negative amounts). Enter each total here and include on your Schedule D, **line 1** (if **Box A** above is checked), **line 2** (if **Box B** above is checked), or **line 3** (if **Box C** above is checked) . ▶							

Note. If you checked Box A above but the basis reported to the IRS was incorrect, enter in column (e) the basis as reported to the IRS, and enter an adjustment in column (g) to correct the basis. See *Column (g)* in the separate instructions for how to figure the amount of the adjustment.

For Paperwork Reduction Act Notice, see your tax return instructions. Cat. No. 37768Z Form **8949** (2012)

Figure 13.3 Federal Tax Form 8949, Sales and Other Dispositions of Capital Assets (page 1)

representing your share of long-term capital gains in the portfolio. Distributed short-term capital gains from these companies are treated as ordinary dividends and are reported on Schedule B (Interest and Dividends).

Dividends are treated differently than ordinary income. An *ordinary dividend* is taxed at ordinary tax rates, that is, your effective

Form 8949 (2012) Attachment Sequence No. **12A** Page **2**

Name(s) shown on return. (Name and SSN or taxpayer identification no. not required if shown on other side.) | **Social security number or taxpayer identification number**

Most brokers issue their own substitute statement instead of using Form 1099-B. They also may provide basis information (usually your cost) to you on the statement even if it is not reported to the IRS. Before you check Box A, B, or C below, determine whether you received any statement(s) and, if so, the transactions for which basis was reported to the IRS. Brokers are required to report basis to the IRS for most stock you bought in 2011 or later.

Part II **Long-Term.** Transactions involving capital assets you held more than one year are long term. For short-term transactions, see page 1.

You *must* check Box A, B, *or* C below. Check only one box. If more than one box applies for your long-term transactions, complete a separate Form 8949, page 2, for each applicable box. If you have more long-term transactions than will fit on this page for one or more of the boxes, complete as many forms with the same box checked as you need.

☐ **(A)** Long-term transactions reported on Form(s) 1099-B showing basis **was** reported to the IRS
☐ **(B)** Long-term transactions reported on Form(s) 1099-B showing basis was **not** reported to the IRS
☐ **(C)** Long-term transactions not reported to you on Form 1099-B

3 (a) Description of property (Example: 100 sh. XYZ Co.)	(b) Date acquired (Mo., day, yr.)	(c) Date sold or disposed (Mo., day, yr.)	(d) Proceeds (sales price) (see instructions)	(e) Cost or other basis. See the **Note** below and see *Column (e)* in the separate instructions	**Adjustment, if any, to gain or loss.** If you enter an amount in column (g), enter a code in column (f). **See the separate instructions.** (f) Code(s) from instructions	(g) Amount of adjustment	(h) **Gain or (loss).** Subtract column (e) from column (d) and combine the result with column (g)
4 Totals. Add the amounts in columns (d), (e), (g), and (h) (subtract negative amounts). Enter each total here and include on your Schedule D, **line 8** (if **Box A** above is checked), **line 9** (if **Box B** above is checked), or **line 10** (if **Box C** above is checked) ▶							

Note. If you checked Box A above but the basis reported to the IRS was incorrect, enter in column (e) the basis as reported to the IRS, and enter an adjustment in column (g) to correct the basis. See *Column (g)* in the separate instructions for how to figure the amount of the adjustment.

Form **8949** (2012)

Figure 13.4 Federal Tax Form 8949, Sales and Other Dispositions of Capital Assets (page 2)

tax rate. Most short-term capital gain distributions from mutual funds are ordinary and are reported as dividends. A *qualified dividend* is one that's subject to the 0% to 15% minimum tax rate on net capital gains. To be a qualified dividend, it must be paid by a US corporation or qualified foreign corporation.[7] A few other rules apply to dividends, including a required holding period for the stock.[8]

TAXES ON OPTIONS

Option profits are taxed in an odd manner. Even if stock is held for more than a year, writing certain types of calls may cancel the benefit of lower taxes on long-term capital gains. Some kinds of options losses cannot be deducted until the other side of a multipart strategy is closed in a future year.

Options trades are normally treated as short-term capital gains and taxed at ordinary rates, even if they are held open for more than one year. When a short call is covered through stock ownership, exercise usually results in the call premium rolled into the net basis, so that the overall transaction is treated as a single short-term or long-term gain or loss.

Key point: Most stand-alone options trades are treated as short-term gains or losses, even if they remain open for more than 12 months.

Covered calls are normally "qualified," which means that upon exercise you are taxed on stock profits and losses under the prevailing capital gains rules. However, if you open an "unqualified" covered call, different rules apply. An unqualified covered call is one written deep in the money, and as long as that call remains open, the period counting up to long-term gains treatment is tolled. This could mean that even owning stock for more than one year could result in treatment as a short-term capital gain. You are allowed to write unqualified covered calls, but you might end up being taxed at short-term rates if the call is exercised.

For example, assume you have owned 100 shares of stock for nine months and the shares have appreciated in value, and you then write an unqualified covered call expiring in 8 months. Five months later, the call is exercised and your stock is called away. Although you owned the stock for 13 months, the gain is treated as short-term. By selling an unqualified covered call, the period counting up to one year was halted and resumed only when the call was closed. However, because it was exercised, the gain never reached the one-year mark.

The level of a qualified call relies on two tests. The first is the dollar value of the underlying stock, and the second is the time until expiration of the option. But in general, an unqualified covered call will be more than one strike or two strikes below current value of the underlying security.

Another set of rules relates to what are called "offsetting" positions. In such positions, a current-year loss is not deductible until the offsetting gain is realized, even if that does not occur until the following year. In this situation, interest payments on margin and transaction fees have to be included in the basis and cannot be deducted separately.

To see a complete explanation of option taxation rules, download the free booklet entitled *Taxes and Investing* from the Chicago Board Options Exchange at www.cboe.com/LearnCenter/pdf/TaxesandInvesting.pdf.

TAX PLANNING BASICS

Profitability based on knowledge of taxes relies on advance planning and timing of transactions. As an investor in the energy sector, you need to be aware of how this year's investment trades are going to affect your tax liability. It may be beneficial in some cases to defer profits to reduce liabilities; in other cases, the greater benefit may be derived from taking losses earlier rather than later.

Tax planning often is based on *tax avoidance*, the planning and timing of trades to maximize benefit and reduce consequences of income taxes. Tax avoidance is completely legal and should not be confused with *tax evasion*, which means falsifying returns or failing to report income. Good tax planning involves calculating outcomes before the end of the tax year and planning transactions based on the current year's estimated taxable income; this means knowing the status of gains as short-term or long-term and taking the current year's income into account when deciding how to time transactions.

The next and final section of the book summarizes energy investing in general by describing common mistakes, myths, and ideas for a successful investment program.

Part IV

Summarizing the Market

CHAPTER 14

TEN MISTAKES INVESTORS AND TRADERS MAKE

> Nothing is more harmful to a new truth than an old error.
>
> Johann Wolfgang von Goethe, *Proverbs in Prose,* 1819

TRADERS MAKE SEVERAL MISTAKES IN HOW THEY SELECT products, when they open or close, and what forms of analysis they use. Following are ten of the most common mistakes investors and traders make.

1. PLACING MONEY INTO TRADES WITHOUT KNOWING THE RISKS

The most common mistake investors make is quite basic: Committing money without awareness of the risks. This means all risks, not just the best-known market risk. With market risk, chances are that after a position is opened, it will lose value. Investors easily fall into the trap of thinking of their basis price as their "zero price," assuming that the value must rise from that point. But the entry price, or basis, is the price in a continuum of rising and falling prices of the auction market. This is obvious, but it needs to be acknowledged.

Key point: The price at which you enter a position is not zero, but only the current level; the next move can be a higher price or a lower price.

In addition to market risk, there are many other types of risk. These include knowledge and experience risk, namely, the risk that an investor will take up a position without adequate understanding of the market, the sector, or the industry. The basic research you are expected to perform requires responsible initial research. Only when you know the fundamental risks (as well as opportunities) can you expect to make an informed decision about where to commit your capital.

Other forms of risk include leverage risk (for those using margin accounts), liquidity risk (selecting investment products that are not easily or efficiently traded), and diversification risk. The latter includes the risk of overdiversification, which is often overlooked because underdiversification is emphasized among financial professionals and in the financial press. You underdiversify when too much capital is placed at risk in a single product or a range of products and all of them are subject to the same risk exposure. For example, while an energy-based ETF provides diversification within the energy sector, it may also be underdiversified if the entire basket of securities is likely to rise or fall based on the same market and economic forces. Overdiversification occurs when capital is spread among so many different products that you can earn only the average of the whole portfolio. This may have been the long-standing problem of the large mutual funds. Because they must broadly spread money among so many companies, they never outperform the market and more often underperform the index of markets.

Another form of risk is the double effect of inflation and taxes. Calculating a breakeven based on these two factors reveals that you might need to earn a much higher rate of return than you think just to maintain net purchasing power. Taxes deplete earnings to the extent of your effective tax rate (for example, if your total federal and state effective tax rate is 40%, you are netting only 60% of your gross investment income). This is not entirely accurate given the special advantages related to long-term capital gains and dividends in some situations, but the point remains valid that taxes reduce your true investment income. Added to this is the effect of inflation. Many people think of inflation as causing higher prices, but another way to look at it is as an eroding force on your capital. For example, if you experience 3% inflation, it means that last year's dollar is only worth 97 cents in purchasing power today. A goal of your investment program may be to attain as a minimum the breakeven rate of return you need to offset inflation and taxes.

The challenge in this is to figure out how to reach that higher gross return without adding risk. There may be solutions, however. The most obvious of these is to focus on stocks yielding high dividends if your portfolio is based on direct ownership of equities. You may further augment income with conservative option strategies, such as covered call writing. The combination of high dividend yields and covered call writing will address the problem of inflation and taxes without adding an unacceptable level of market risk.

2. Making Decisions Based on Advice Instead of Research

Do you make your own investing decisions or do you rely on a professional advisor? Many people benefit greatly from hiring an experienced and knowledgeable professional. However, the degree of reliance on advice determines how effectively it can be used to find risk-appropriate products and trade them.

It could be a mistake to rely completely on the advice of others, even that of competent professionals. No one is going to take care of your financial interests as well as you will. This does not mean a financial advisor has to be proficient in the energy sector. It does mean that the professional needs to comprehend a full range of risk exposure, articulate your own risk tolerance level, and match your risk tolerance to appropriate products.

The most effective use of an advisor then is to rely on recommendations to the extent that outside research is put to use. At the same time, if you perform your own research and then consult with the financial advisor, you are fully involved in making your decisions. The very idea that someone can be paid to give you sound advice is dangerous. You certainly can get such advice, but there are no guarantees that it will be good advice for you.

> **Tip:** To discover whether a financial advisor is right for you, insist on being involved from the start. The advisor's reaction to this demand reveals all you need to know.

This problem of finding an effective advisor also has to be analyzed in the context of how that individual is compensated. If your advisor works on commission, of course, you will be steered toward commission-paying products, such as load mutual funds. These are not likely to outperform no-load funds, but your advisor is motivated to get paid for the services provided to you. The built-in conflict of interest is impossible to ignore. If you pay an advisory fee, you are likely to get far more objective advice, including consultation about ideas you have developed based on your own research. An advisor who does not want you to perform your own analysis should be viewed with skepticism; you should be involved in managing your own money and not left out of the analytical side of finding sound, appropriate products. If an advisor is paid via a fee *and* also gets commissions for products recommended to you, then you have the worst of both worlds. How can an advisor be objective when commissions are involved? It is possible, of course, but less likely. When it comes to your finances, the more clarity you can achieve, the more confidence you will have in your decisions.

3. Limiting Research to a Short List of Indicators

Most fundamental investors have to limit the sheer number of indicators they rely upon, or they may become lost in an array of dozens of ratios and trends. Using as many indicators as you can does not necessarily increase the value of the end result. It is a mistake to try and find value in an excess of numbers. However, it is also a mistake to use too small a number as basis for your decision.

The "right" number of indicators should be based on the type of information you gather to make your decision to buy one product over another. For example, your list could include a few strong and reliable trends in revenue and earnings, dividend yield, P/E ratio, and debt ratio. This list of only four indicators provides information about growth and profitability, current yield, price based on earnings per share, and capitalization (equity versus debt). You can further expand and analyze the net return represented by earnings and the payout ratio represented by dividends.

Tip: You do not improve the quality or value of information by increasing the number of ratios or trends you follow. You're better off with a very short list of indicators that tell you all you need.

The point is that if you use too many indicators, you lose effectiveness. However, if you rely on too few, you create the opposite problem. You need to track earnings growth, dividends, price/earnings, and debt to fully appreciate whether a company is gaining or losing strength within its sector. At the very least, you will need to cover the basics of earnings growth, price level, and working capital.

You can apply the same rule to technical analysis. Studying charts, you can apply dozens of indicators to try to time entry and exit, but using too few or too many limits the value of what is revealed. A key to timing is that reversal indicators need to be confirmed. For this, you need reliable signals, but relying on only one or two may cloak developing trends. On the technical side, you need to track price patterns, volume, and momentum.

Price patterns come in many shapes and sizes based on either Western or Eastern signals, but some are stronger than others. While you may recognize many evolving signals, you could decide, for example, to focus on three Western signals, such as breakouts, price gaps, and double tops and bottoms as well as on three Eastern signals, such as engulfing patterns, white soldiers or black crows, and hammer or hanging man. For volume analysis, you may select recognition of volume spikes, On Balance Volume (OBV), or Chaikin Money Flow (CMF). And for momentum you may track Relative Strength Index (RSI) or Moving Average Convergence Divergence (MACD).

On both the fundamental and technical sides, there are dozens of additional signals you can check, and relying on too many is as flawed as using too few. In fact, the most effective form of analysis is to combine four to five fundamental indicators with four or five technical indicators and round out the program by analyzing these together.

4. Forgetting the Importance of Industry Research

With so much emphasis on well-known fundamental analysis (based on the ratios and trends found in the financial statements), it is easy to overlook the equally important, broader version of fundamental analysis, based on the specific industry.

For example, industry-specific fundamental analysis in the retail sector is influenced by the overall health of the economy and buying trends; the gaming and leisure sector also is influenced by how consumers feel about the economy, with more leisure activity taking place when the economy is strong. The energy sector has its own, quite unique set of fundamentals beyond balance sheets and income statements.

Key point: The fundamentals of the energy sector, whether economic or political, define the investment opportunities within the sector. This, coupled with company-specific tests, helps you to define the investment value of an energy stock, ETF, or futures position.

In fact, the energy sector is the only market sector that is influenced by economic and geopolitical factors, some real and some perceived. If consumers believe that oil and gas are running out, and that oil is damaging the ozone layer, or that OPEC is going to raise prices, these perceived "facts" influence prices. Whether they are true or not doesn't matter. In the energy sector, perception is as important as proven fact. To a degree, this phenomenon of perception having as much weight as actual fact affects all markets and sectors. However, in the energy sector, it is a *prevailing* market force. Perception has greater weight than actual economic forces of supply and demand.

For energy, actual and factual realities often are quite different than perception. The truth is that new reserves and new extraction methods are announced regularly, and it now appears that supplies are plentiful in the United States, especially the supplies of natural gas. The question of whether fossil fuels are causing or aggravating global climate change has not yet been settled, and, in fact, there is no proof that global warming is anything other than a normal and cyclical occurrence. As one opinion article explained this controversy and the consequences for those who dare to challenge the believers in global warming:

> Perhaps the most inconvenient fact is the lack of global warming for well over 10 years now. This is known to the warming establishment, as one can see from the 2009 "Climategate" email of climate scientist Kevin Trenberth: "The fact is that we

can't account for the lack of warming at the moment, and it is a travesty that we can't." . . .

Although the number of publicly dissenting scientists is growing, many young scientists furtively say that while they also have serious doubts about the global-warming message, they are afraid to speak up for fear of not being promoted—or worse. They have good reason to worry. In 2003, Dr. Chris de Freitas, the editor of the journal *Climate Research*, dared to publish a peer-reviewed article with the politically incorrect (but factually correct) conclusion that the recent warming is not unusual in the context of climate changes over the past thousand years. The international warming establishment quickly mounted a determined campaign to have Dr. de Freitas removed from his editorial job and fired from his university position.[1]

It does not matter what position anyone holds regarding these issues; the fundamentals in the energy sector must include more than the profit and loss or capitalization of an energy company.

Research ultimately wins arguments, but it may take years to refute the arguments against expanding energy production. The perceptions and political motives involved are difficult to overcome, but as findings are made available on the Internet, a more informed public will eventually communicate with elected officials and create a dialogue leading to the right answers. The outcome of this debate is likely to set up a balance between environmental sensitivity and satisfaction of the demand for oil and gas products, but this will take time. Meanwhile, investors cannot restrict their fundamental analysis to the well-known financial trends. As an investor, you need to also be aware of the perceptions of economic and geopolitical realities in the energy market.

5. Going for Fast, Small Profits Rather than Longer-Term Growth

Swing traders and day traders move in and out of positions by design. Their purpose is to take a high volume of small profits, with the idea that this is preferable to waiting out price movement. Some subscribe to this approach, and others do not. But if you are a believer in seeking value and growth with more safety, you probably are more likely to act as a buy-and-hold investor rather than as a short-term trader.

Longer-term growth has to be managed, however. Even the most conservative investor has to decide when or if to take profits when they appear. It makes sense to set goals in advance for when profits will be taken, and if you can stick with your own self-imposed rules, you will probably accomplish your goals. However, what happens when a position does not reach the profit goal?

> **Tip:** Decide whether you prefer buy-and-hold strategies or short-term, smaller profits. Neither is right or wrong, but you do need to decide in order to set up your trading and investing program.

If your position does not reach your goal, you need to have a second price level, a bail-out point. This is the price at which to close out a position to minimize losses. For example, you might buy stock in an energy company and set two price levels: Sell when you make 25% profit or when the value declines by 10%. These percentages are arbitrary and should be varied based on dollar amounts, your risk tolerance, and what you realistically expect to earn (or lose) on investments.

Another portfolio management skill is to take steps to avoid having loss positions on hand. If you take profits when they materialize but never take losses, you will end up, by attrition, with a portfolio of positions below your basis. Most investors who have taken profits as they appeared know exactly how this works. You sell your successful positions and take profits, but keep the unsuccessful ones. This is an upside-down strategy. The solution: When you are able to take profits, you should also close out a position and accept a loss. This makes sense because otherwise you may end up keeping capital committed to positions that might never turn profitable, or if they do, this could take too long. By closing to take losses along with profits, you achieve several benefits.

First, you reduce tax liabilities in the current year because the losses offset your gains. Second, you dispose of loss positions in your portfolio, freeing up more capital to invest elsewhere. Third, you avoid the possibility of a small loss becoming worse and turning into a larger loss.

No matter how you invest or trade (short-term versus long-term), taking profits and losses is going to work out more successfully when you work within a plan. Set profit and bail-out points on the basis of dollar values or percentages. If you hold equities in your portfolio, losses can also be mitigated by using stop-loss orders. This is a type of order that automatically generates a sale when the stock reaches a predetermined level. For example, you may set the stop-loss at 10% below your cost. However, even though the sale occurs once that price level has been reached, it does not mean the sale takes place at that price. If the price continues to fall, the actual sale could be at a lower level. Stop-loss orders are a good way to limit your loss exposure, but they do not guarantee the sale price.

6. Failing to Follow Your Own Rules for Exiting Positions

Setting rules is easy compared to *following* rules. Many investors set profit and bail-out points but then do not enforce those limits. When the profit level is reached, it is tempting to hold off in the hope that profits will move even higher; they might, or they might evaporate. But if you abandon your action point, when do you sell? The same applies when a loss takes place. Instead of bailing out, some investors and traders rationalize that the price might turn around and go back to "zero," so it makes sense to hold off. But if losses remain or get worse, when do you sell and take your loss?

Key point: Setting rules and goals is the easy part. It gets more difficult when the time comes to *follow* your own rules. This is where most goals unravel.

Failing to follow your rules for exiting a position is a big problem for investors. In fact, exiting is more difficult than entering a position because timing is everything. Remember, investors tend to think of their entry price as the zero point, with the expectation that profits will begin to accrue right away. This is probably true only half of the time. The initial assumption—that entry price is effectively the zero point—is only the first problem in how investors trade. A more serious question is when to exit—not just setting the goal, but adhering to it.

A very widespread belief among investors is that they should focus on finding exceptional bargains to purchase. However, even if you do find those, when do you close the position, if at all? As long as a stock is performing well (and perhaps this means it stays at about the same price as you bought it and is not losing value), holding for the long term and collecting dividends is a perfectly acceptable strategy. In such a low-volatility investment, adding in covered calls or similar options strategies further mitigates market risk and increases current income.

Investors may find and stick with profitable buy-and-hold strategies, especially when focused on dividend yield and covered calls in addition to stock ownership. But a related problem occurs when a self-described "conservative" investor acts like a speculator. If a trader adheres to conservative principles but moves in and out of trades often, is this truly a conservative strategy? Does the action follow the self-imposed rule?

As part of the rule for when to exit a position, the profit goal and loss bail-out should be set realistically to avoid the mistake of rapid movement of funds in and out of positions. A short-term swing trading strategy is one of several ways to trade in the market, but only if that is the purpose for trading. Problems arise when traders think of themselves as conservative and focus on buy-and-hold strategies, but end up taking small profits as soon as they materialize. They are left with the dilemma of needing to invest funds elsewhere, a next step that is often overlooked. As part of the impulse to trade often and to take even the smallest of profits, the same traders may easily fail to have a follow-up strategy in mind. Consequently, they often invest funds without the requisite research and analysis.

If you are a short-term trader, these issues do not matter because you should have an unlimited supply of potential trades you can make at any time. But if you think you are a conservative long-term investor, make sure your trading patterns confirm that description.

7. Misunderstanding How Diversification Works

The concept of diversification is well understood, but its practice is not. Buying several energy stocks is not effective diversification because all are subject to the same forces of supply and demand and will tend to move with the market in the same direction. In comparison, diversifying among energy and other sectors is far safer because it spreads risks not only among different companies but also among different sectors.

The diversification accomplished by buying shares of an ETF is a questionable system Though It is promoted for liquidity and diversification. ETF shares are very liquid because they can be traded just like shares of other equities, but ETFs are not diversified. They contain a basket of securities with similar features. Investors who want to be in the energy market but are not certain about which companies to buy will be wise to invest in an ETF—not for diversification but to spread risks among many similar corporations in the same sector.

Tip: Diversification does not mean buying shares of many different companies sharing the same market vulnerabilities. It could, in fact, mean doing less rather than more.

True diversification will involve spreading risks among different sectors as well as among different products. For example, stock market risks are diversified by investing some capital outside of stocks. This type of diversification, also known as *asset allocation*, assigns percentages of total capital to equity, debt, real estate, currencies, and other classes of assets.

Another question has to be asked with an eye on the danger of both under- and overdiversification. Is diversification necessary? This question may seem shocking because it questions a widely assumed premise: that you have to diversify your holdings to reduce risks. But for some investors, diversification will only reduce the overall profits from a strong portfolio. With the use of hedging attributes found in options (as well as the natural hedge between stocks and futures), it might be possible to guard against risk without diversification. For example, if you are a committed energy investor, what does it mean if you own energy stocks, debt securities, futures, ETFs, and commodity index funds? It means you are not diversified in the sense of the sector in which you have invested. But the question should be raised: Do you need to diversify if you have hedged risk by spreading capital among different products within the same sector?

In other words, energy investors who invest in a range of different products—all within the energy sector—are diversifying under some definitions. If you believe that energy value is going to rise inevitably, then why force yourself to also buy hotel, gaming, retail, and manufacturing stocks? The issue of diversification is complicated because you can underdiversify or overdiversify, or diversification itself can reduce profitability. This is why the question of whether you need diversification is especially difficult regarding the energy market. Most investors should diversify between energy and other markets and between stocks and other products, but if you want to focus exclusively in the energy market, consider looking beyond stocks and diversifying within the energy sector in other ways.

The perceived advantages of diversification are easily offset by the cost and complexity of overdiversification. Adding that "safety net" to your portfolio does not make it easier to manager but can make this task more difficult:

> Many investors may incorrectly assume that having a diversified portfolio means they can be less active with their investments. The idea here is that having a basket of funds or assets enables a more laissez-faire approach, since risk is being managed through diversification. This can be true, but isn't always the case.[2]

8. Forgetting to Combine Fundamental and Technical Indicators

You might tend to favor fundamental over technical analysis or technical over fundamental. Both offer strengths and advantages, but both also are only part of the total picture. Using both to pick stocks and to determine entry and exit makes sense. Using both also provides a means for determining when the strength of a company has changed, that is, when yesterday's strong "hold" candidate becomes a "sell" position.

Key point: Analysis is not limited to deciding what to buy. It is equally effective in identifying when things have changed and when it is time to sell.

Fundamentals are backward-looking in the sense that they report on trends over several periods and indicate relative strength or weakness. Even so, that backward look provides a basis for estimating the future trend as well. The method is not flawless, but it provides the best information available for judging a company's competitive and market strength, earnings potential, and management of working capital.

Technical analysis is focused on charting and the study of price, volume, and momentum. Chartists rely on technical signals to time short-term entry and exit and tend to accept a higher level of risk than the more conservative fundamental investors. This distinction (between traders and investors) often defines the related distinction between high risk and low risk. However, this does not have to be the only way to use both fundamental and technical analysis. In fact, the most conservative means for picking products is to use both fundamental and technical indicators.

For both types of analysis, using a very limited number of indicators is not only smart but also essential. If you rely on too many indicators, you do not improve the quality of information, and the decision point could then be more difficult to reach. For example, on the fundamental side, you can gain excellent insight about companies and make valid comparisons with four key indicators (revenue/earnings, dividend yield, P/E ratio, and debt ratio). This does not mean these are the only indicators you should use, but it may form the foundation of an analytical program. On the technical side, you may be aware of dozens of reversal indicators but focus on only a few (for example, two of each for price, volume, and momentum). There are so many fundamental and technical indicators to choose from that you have the ability to make your program work flexibly.

In the energy sector, for example, you may select charts and automatically add favored technical indicators to the price track, so that any changes in price, volume, and momentum are visible immediately. On the fundamental side, you can narrow down your list with comparisons of the four fundamental trends and indicators. Together, these two methods help you to find and use a wealth of information in a short list of tests you apply to your selection of investments.

9. Ignoring the Importance of Confirmation and Proximity

On the technical side of the equation, finding powerful reversal signals is invariably the key to timing. A reversal signal forewarns you that it is time to get into a position or to close out one you are in already. A continuation signal, although not used as often, also tells you that the current trend is most likely to continue and that the time to make that move is not here yet.

The biggest problem with these signals is not recognition, but confirmation. Finding good signals is not difficult; in fact, reversal signals pop up all the time. But there are two aspects to the timing of trades that are often overlooked: confirmation and proximity.

Confirmation is the finding of a second signal predicting the same price change as the first. That is, a reversal signal is found and then confirmed. This may be in the form of Western indicators (double tops or bottoms, head and shoulders, large price gaps, triangles, or wedges, to name a few), Eastern or candlestick indicators (there are dozens of candlestick reversals, but some are stronger than others—such as the engulfing pattern, which is found often and is reliable as a reversal indicator most the time), volume indicators and spikes, and momentum oscillators.

Failing to find confirmation is a big mistake, because no signal by itself is reliable enough to generate a trade decision. If confirmation is not found, you should not act. If a confirming indicator contradicts the initial reversal, you should consider this another signal for not taking any action.

Proximity describes the location of current price in comparison to resistance and support. A reversal signal is strongest when it occurs close to these trading range borders. If price moves through these levels, notably with price gaps, the odds of reversal are very high. If you see a reversal signal at the resistance or support level and then find confirmation, the likelihood of reversal is greater than for any other proximity level within the trading range. The entire study of technical analysis is based on price patterns, specifically on failed trends at and even beyond resistance and support.

> **Tip:** Confirmation and proximity are both essential. But when they are combined, your ability to time trades is vastly improved.

10. Avoiding Markets Based on Misperceptions

One attribute of the stock market is that it functions more as a rumor mill than as a vehicle for facts. In addition to the daily and widespread creation of rumors, perceptions—true and false—persist in the market. In the energy market, myths and misperceptions are probably more widespread than in any other sector.

The beliefs that oil is about to run out, that it creates global warming, or that alternative fuel sources will work just as efficiently and as economically are only a few examples of such perceptions. They *might* be true in part, but there is no science yet to settle the questions. As a result, many investors make decisions (in energy and in other sectors) that are based not on economic facts, but on economic perceptions.

Key point: It might be politically incorrect to question the perceptions about energy. But that does not change the facts, whether they are generally known or not.

An example from recent years makes this point. Ethanol was promoted as a cheap, widely available, and clean fuel alternative to gasoline. But it is not cheap; the effects on engines may prove to be quite high: "In smaller engines, ethanol can create a chain reaction of events that ends up clogging valves and rusting out small metal parts—including, crucially, carburetors."[3]

Food prices have also been affected adversely by increased focus on agricultural crops for ethanol rather than for food products. As long ago as 2008, in fact, "researchers at Purdue University in Indiana found that corn prices had risen to $4 a bushel, the highest in a decade, largely because of the higher prices farmers can demand from fuel producers."[4]

The misperceptions about energy sources create many problems but also many opportunities. Investors seek to find bargain prices coupled with high demand. The energy sector meets this standard elegantly. Demand is not going to disappear, and the true science and economics of supply and demand make the energy sector a very desirable market for investments. The product is a necessity, but its reputation has been maligned for so long that many untruths prevail. The next chapter lists and summarizes these misperceptions and myths.

Chapter 15

Ten Myths of the Energy Industry and Futures Market

To die for an idea is to place a pretty high price upon conjectures.
Anatole France, *La Révolte des Anges*, 1914

In the energy sector, myths and misperceptions are more powerful than science, economic truth, or observable facts. All other market sectors are based on the economic realities—supply and demand—and on the company-specific news published concerning financial health, potential mergers, new product announcements, and earnings, of course.

Some perceptions are true, but many are controversial or false. In the energy sector, there are at least 10 myths that have greater influence over investment potential than actual facts do.

1. Oil Will Run out in a Few Years

For over 100 years, predictions have been offered about the impending demise of oil as a resource. People and agencies that make predictions often have two attributes worth mentioning. First, they may have an agenda in making the prediction. For example, an anti-oil and pro-alternative energy agenda is one of the motives for predicting that in the near future there will be no oil. The implication, of course, is that if we don't develop alternative energy sources now, we will suffer once the reserves have simply dried up.

The second attribute is that making predictions of a pessimistic nature is a very human tendency, and those who predict a dire future often take pleasure in making the prediction. There is little publicity in predicting that all will be well and happy in the future, but headlines are created when the word is put out that in 10 years there will be no oil.

Key point: In spite of predictions over the past 100 years that world oil supplies are about to disappear, new reserves and new extraction technology contradict all of the predictions. As of 2013, production is exceeding demand.

The facts are quite different from the predictions; in fact, the peak oil theory has fallen into disfavor among industry experts. Now people who used to make predictions about the timing of peak oil claim that the peak has already occurred, even while production numbers contradict that claim. A 2013 article cited an assessment by the International Energy Agency (IEA) predicting that

> the surge of supply from North America—most of it from new unconventional sources—will transform the global supply of oil and help ease tight markets. Between now and 2018, the IEA projects that global oil production capacity will grow by 8.4 million barrels a day—significantly faster than demand. Oil isn't likely to peak any time soon. And that's bad news for climate policy.[1]

The conflict in this myth is between the agenda of climate policy, which is based in part on the belief that oil is a major cause of climate change, and the simple demand factors. Added to this is the reality that production capacity currently outpaces demand.

Investors are the beneficiaries of this myth, because as long as the myth of scarcity holds, investment value continues to rise. The irony here is that the economic *facts* tell a different story. Production is outpacing demand, and reserves appear plentiful for the coming few decades at least. If natural gas is added into the equation of supply, the United States alone has over a century of supply based on current demand.

2. Most Oil Is Found in the Middle East

Ask anyone where the United States gets most of its oil. A large number—perhaps most—people believe that most of the oil consumed in the United States comes from the Middle East. In fact, the United States produces 39% of its own oil, and another 35% comes from Canada and Mexico. Only 13% is imported from the Middle East (and another 13% from other countries).[2]

Key point: Oil used in the United States comes mostly from North America. The country's interests in the Middle East go beyond the need for oil.

This means that within a few years the United States could be self-sufficient and would not need to rely on the Middle East and specifically, on OPEC, for oil imports. This will require development of known reserves offshore and on land, including the Alaskan fields where reserves are vast. Alaskan oil is a major source of total US oil produced and refined: "Alaska's crude oil production peaked in 1988 at about 738 million barrels, which was equal to about 25% of total US oil production. In 2011, it was about 209 million barrels, or about 10% of total US production."[3]

A long-standing controversy concerns the Arctic National Wildlife Refuge (ANWR), a Northern Slope area of Alaska that is relatively small but containing oil reserves estimated to be very large. ANWR involves only 2,000 acres of drilling area out of 19 million acres in total in the ANWR region. Known reserves in the 2,000 acres could replace imports from Saudi Arabia for the next 20 years, according to an article written by a former governor of Alaska.[4]

Most of the oil used in the United States is derived domestically and from North American trade partners; however, the myth of oil from the Middle East persists. The interests of the United States in the Middle East go beyond oil and involve geopolitical motives as well as natural resources; the issue is clouded not only by the myth of dependence on Middle East oil, but also by the unknown levels of oil's negative role in matters of environmental and climate change.

3. Sand and Shale Oil Are Too Expensive to Extract

The cost of extracting oil from sand and shale reserves has gone down in recent years due to improved technology. The issue of environmental problems remains a serious one, since oil extraction from these two sources of negatively impacts the areas where mining takes place.

> **Tip:** Oil reserves are not always accessible; the price has to be high enough to justify the cost of extraction, and as technology improves, more known reserves become available.

Oil sands (also called tar sands) are a combination of clay, sand, water, and bitumen, a form of heavy oil. Bitumen is refined into oil; however, extraction is usually accomplished through strip mining or open pit mining, both of which are undesirable in terms of environmental effects. The reserves of oil sands provide an incentive to continue searching for alternative mining techniques, however. An estimated two trillion barrels exist in the world, including between 12 and 19 billion barrels in Eastern Utah.[5]

Oil shale is found in sedimentary rock containing bituminous materials. Extraction requires heating and separation of oil from the rock. Large deposits are located in the Green River formation in Colorado, Utah, and Wyoming, holding an estimated 1.2 to 1.8 trillion barrels of oil; recoverable levels are about 800 billion barrels, equal to all of the proven reserves in Saudi Arabia. This is enough to satisfy US demand at current levels for over 400 years.[6]

The process of bitumen recovery requires a large volume of water and thus means an environmentally negative impact for sand and shale oil. In Alberta, Canada, alone, 359 million cubic meters (m^3) of water are used per year to mine bitumen, and only 10% of that water is returned to the water supply.[7]

Considering that this is only one localized example in a province of Canada, the global environmental impacts associated with sand and shale oil are enormous. The reserves are vast, but improved methods of extraction are necessary before they can be safely mined. An alternative to the environmentally damaging and expensive older methods may be found in *fracking* (hydraulic fracturing). In this method, rock layers are fractured with high-pressure insertion of water. This releases the oil or natural gas, which can then be extracted. This procedure is controversial, but with improved environmental safety features, fracking is likely to provide future benefits to access the vast oil and gas reserves in sand and shale.

4. Prices Are Driven Mostly by Speculation

The belief that speculation is the cause of prices, especially in futures, is irrational, but like so many other myths, it persists. Many traders speculate in stock prices as well, both through stock and option trading, but they are not accused of manipulating stock prices or of causing a rise in prices. But when it comes to futures, many financial journalists, insiders, and other "experts" choose to believe that oil prices are high due to speculation.

> **Key point:** There is something satisfying about simplifying the pricing problem by blaming it all on speculators. The facts contradict this claim, but this is an example of perception holding greater weight than fact.

Fox News personality Bill O'Reilly stated this opinion and was cited in an article about the matter during a debate with John Stossel:

> O'Reilly suggested oil speculators are gambling "crooks" and that there should be a required 50% fee on their speculation. "They don't want the oil. If they want to go to Vegas, go. If they want to get to Mohegan Sun, go. We all need oil to live, and I don't want a commodity that my family needs to live to be the subject of speculators who are gambling like it would be in Vegas."[8]

The claim is unsupported. In 2008, the Commodities Futures Trading Commission (CFTC) studied the effects of speculation on oil prices and concluded that speculation was not the cause of a run-up in prices, but a reaction to price changes caused by movement in supply and demand:

> The Task Force's preliminary assessment is that current oil prices and the increase in oil prices between January 2003 and June 2008 are largely due to fundamental supply and demand factors. During this same period, activity on the crude oil futures market—as measured by the number of contracts outstanding, trading activity, and the number of traders—has increased significantly. While these increases broadly coincided with the run-up in crude oil prices, the Task Force's preliminary analysis to date does not support the proposition that speculative activity has systematically driven changes in oil prices... changes in futures market participation by speculators have not systematically preceded price changes. On the contrary, most speculative traders typically alter their positions following price changes, suggesting that they are responding to new information—just as one would expect in an efficiently operating market.[9]

The stock market has had periodic price bubbles and panics, caused in part by the herd mentality of the market; however, these tend to be short-term and prices return to the trend of the longer term and to the primary trend of the market. One factor often overlooked by those blaming speculators is that the futures market—unlike the stock market—is a zero-sum gain market. In the stock market, capital value builds as companies profit, but in futures, one trader's gains are always offset by another trader's losses. Either the buyer or the seller wins depending on the direction in which the price moves. But in a zero-sum gain market, manipulation of the underlying price is impossible.[10]

5. Alternative Energy Will Solve All of our Energy Problems

Worldwide, most energy is derived from petroleum, natural gas, and coal; only a very small portion comes from renewable sources. This is a reality, and if any changes are to occur in the balance between fossil fuels and other sources, it will take decades for them to come about.[11]

Alternative energy usage has not grown in recent years; in fact, coal continues to be the fastest-growing energy source worldwide:

> Fossil fuels still dominate energy consumption, with a market share of 87%.Renewable energy continues to gain but today accounts for only 2% of energy consumption globally. Meanwhile, the fossil fuel mix is changing as well. Oil, still the leading fuel, has lost market share for 12 consecutive years. Coal was once again the fastest growing fossil fuel, with predictable consequences for carbon emissions.[12]

A problem with alternative energy is that it is expensive to produce and the yield benefits rarely outweigh the costs. Some forms of alternative energy also have unintended consequences. For example, wind power costs seem to rise each year even as wind turbine farm construction grows. A technical report sponsored by the U.S. Department of Energy concluded that

> from roughly 2004 to 2009, the cost of wind energy increased. Historically, cost reductions have resulted from both capital cost reductions and increased performance. From 2004 to 2009, however, continued performance increases were not enough to offset the sizable increase in capital costs that were driven by turbine upscaling, increases in materials prices, energy prices, labor costs, manufacturer profitability, and—in some markets—exchange rate movements.[13]

Among the unintended consequences of production of alternative energy are environmental concerns. Birds, including rare birds such as the golden eagle, often fly into wind turbines and are killed. This creates a bizarre paradox in the environmental community where protected species are protected by the law and where alternative energy is touted as the way to save the planet. As one report on this issue stated regarding deaths of rare birds:

> Killing these iconic birds is not just an irreplaceable loss for a vulnerable species. It's also a federal crime, a charge that the Obama administration has used to prosecute oil companies when birds drown in their waste pits and power companies when birds are electrocuted by their power lines.

> But the administration has never fined or prosecuted a wind-energy company, even those that flout the law repeatedly. Instead, the government is shielding the industry from liability and helping keep the scope of the deaths secret.[14]

The article also cites a source (the *Wildlife Society Bulletin*, at http://joomla.wildlife.org) that claims over half a million birds are killed each year in collisions with wind turbines.

Similar problems are associated with virtually all forms of energy, including nuclear, electric, and of course, oil and gas. However, the claim that alternative energy solves all energy problems is not accurate; alternative energy involves secondary negative impacts. The costs of alternative energy continue to make many of the favorite sources (solar and wind among them) impractical on a cost-benefit basis, but the myth persists that these alternatives are the solutions of the future.

Key point: The desire for clean, cheap energy is universal, but the reality is that finding such energy is going to require decades of research. Meanwhile, it is difficult to argue with the efficiency and availability of fossil fuels.

In some applications, alternative energy sources have been shown to be more expensive than the fossil fuels they are meant to replace. For example, in spite of widespread promotions for the efficiency and cost benefits of solar energy, some users have discovered that the costs are higher than they had expected. In one New Jersey school district, for example,

> the lure of solar panels to produce renewable energy that would pay for itself proved very tempting, but some who took the plunge are now finding the cost benefits are shrinking. . . . But in a presentation last month, officials learned that instead of paying for themselves, the solar panels have become a liability. Now, about 20 percent of the money the district spends on paying off its accumulated debt is going toward paying for the solar installation.[15]

The reality is that fossil fuels, even with their impact on the environment, are efficient and profitable in use, especially compared to wind and solar energy and other alternative energy sources. Until cheaper and more cost-effective methods are developed, alternative energy does not pose a realistic solution for future energy requirements.

6. Balancing Energy Demand with Climate Issues Is Impossible

The conflict between the energy industry and environmental agencies in the government (as well as private businesses or special interests) has been going on for many years. The problem is that there often is no middle ground between the two sides. Some believe that expanding drilling and exploration of fossil fuels is good no matter where it takes place; others believe that all fossil fuel should be replaced with alternative fuel sources and that it is a matter of saving the planet from the consequences of climate change.

> **Tip:** To enter into a dialogue about between energy and environment, both sides have to be willing to seek compromise. Over the past few years, a dialogue between the two sides has been evolving, a positive trend.

However, oil sector companies spent over $35 billion on environmental research and health initiatives in 2011, and by 2030 that figure will rise to an estimated $56 billion. Efforts include improved detection of pipeline leaks and better oil spill containment methods; oil companies are increasingly working in partnership with environmental interests.[16]

The issue was summarized in an article by two experts, a professor of petroleum engineering and a geologist:

> Over the decades of petroleum extraction, the industry as a whole has done an enormous amount to balance the production of resources with minimal environmental impact. There is, and always will be, a dynamic tension between corporate profits and public good; however, mature corporations have implemented policies and practices that recognize this balance. This is and will remain a constantly moving target. Instilling

> and maintaining corporate philosophies that encourage and reward practices that reduce environmental impact is key.[17]

As long as industry insiders continue working with environmental interests and promoting research into improved methods, progress will be made, and the balance between energy demand and environmental interests is likely to continue to improve. The environmental sector has provided a true service in requiring energy companies to take steps to reduce negative impacts. This applies not only to emissions standards, but also to extraction and transportation policies across the oil and gas supply chain.

7. Futures Are Best Traded Directly

The futures market has been active in the United States since the 1840s, when Chicago was the hub for buying and selling agricultural commodities. The perception of direct trading in futures is that it is specialized and that considerable experience is required for anyone trading in this market.

The perception is correct in the sense that direct trading in futures demands considerable capital. Even on margin, buying or selling single futures contracts may involve huge amounts of capital, depending on the value of the contract. For example, ethanol futures trade in increments of 29,000 gallons; Brent crude oil in 1,000 barrels, and heating oil in 42,000 gallons. The dollar value of a single contract is quite high, although margin is much less. In the summer of 2013, a heating oil initial margin requirement for one futures contract, for example, was approximately $12,000.

> **Tip:** Trading individual futures contracts can be very expensive, even based on margin. Any use of leverage—such as margin—increases both opportunity *and* risk.

Buying or selling a single futures contract at these levels is expensive, and with single contracts it is impossible to diversify effectively. However, you can trade futures through other products. These include options on futures, energy ETFs, or commodity futures index funds.

Table 15.1 Energy portion of commodity fund holdings

Commodity Index Fund	Energy
S&P World Commodity Fund	73.4%
Goldman Sachs Commodity Index (GSCI)	69.7
Rogers International Commodity Index (RICI)	44.0
Reuters/Jefferies CRB Index	39.0
Dow Jones UBS Commodity Index	37.6

The index fund is very attractive because it is an index of a great number of futures, dominated by energy futures. The five best-known commodity index funds report as shown in Table 15.1.[18]

Focusing more on energy than on other commodity sectors, the index fund alternative is probably the most efficient and economical method for investing in this market. The diversification into other commodities not in the energy sector is healthy diversification, providing investors with a means for reducing risks and gaining profits from worldwide trends in commodity prices.

8. The Futures Market Always Carries High Risk

Direct purchase or sale of futures contracts may be defined as carrying high risk if only because the contract price is so high for many energy futures and also because they are highly leveraged. Whenever leverage is involved in a product, risks are also increased. With leverage, profits may be gained more rapidly, but losses are also accelerated.

This has kept many investors out of the futures market. Indeed, anyone intent on investing in energy is more likely to trade energy stocks or ETFs. However, the availability of commodity index funds and options means investors can find alternatives and pursue them while also minimizing their risks.

Key point: You can invest in energy futures in many ways, and some carry a lower risk compared to direct transaction of energy futures.

The risk issue should be central to the choice of futures as well as to any other investment decision. Risk comes in many forms other than market risk. When investors avoid the futures market due to awareness of high risk in direct trading of futures contracts, they can easily overlook other risks, which could be even more serious. Even those buying and holding shares of stock have to be concerned with market risk and must carefully weigh their knowledge and experience, diversification, leverage, liquidity, and risk involving lost opportunity, inflation, and taxes. It is not realistic to focus on any one form of risk or to reject a market based on a narrow focus.

Risk also has to be assessed based on the type of product involved. The futures market is dominated by the energy and agriculture sectors, both producing necessities of life. For this reason alone, there will always be a market for energy and agricultural commodities. The risk question should be: How do you access that market? Trading in necessities makes sense on a purely economic basis, which is not always the dominant consideration in determining investment value. However, if you believe that economics eventually win out, it is clear that the energy sector will always combine high demand with limited supply (real or perceived). The question of access goes back to choices of products within the sector: equities, debts, ETFs, index funds, options, or direct trading of futures contracts. The risks associated with direct trading in futures are higher and more serious than those connected with any other products and avenues. However, it is a mistake to label futures trading a high-risk endeavor without further understanding that the risk is not in the futures themselves but in the method of involvement. To trade futures directly, you need experience and knowledge about delivery risks. Furthermore, the incremental value of a futures contract is likely to be far too high for most traders, as it is in the five-figure range for many energy futures. Finally, the cost of transacting a single futures contract is going to be much higher than the equivalent cost of an investment in an ETF, commodity fund, or option.

The same reasoning is applied as well to other markets. Some investors avoid the stock market because it carries high risk—perhaps because they have lost money in stocks in the past. However, was the loss due to the stock market being always a high risk, or was it due the individual investor's poor timing or poor choice of high-risk stocks? For some, the solution is to apply fundamental and technical analysis, but this will not mitigate all forms of risk either.

The bond market, ETFs, commodity index funds, and options markets all carry specific risks. However, these are not so much caused by high risks inherent in futures as by the risks inherent in the particular product. Based on timing, any market may result in losses and may consequently be viewed as a high-risk market. The danger of this conclusion is that you might end up in a market with even higher risk by trying to avoid one perceived as having a high risk, when in fact on a relative basis its risk is not that high. The following are useful guidelines for risk analysis of the futures (or any other) market:

1. Understand the attributes of the market and how it functions.
2. Study all forms of risks as they apply to the market and to the types of products involved.
3. Compare the risk attributes of the market by product type (stocks, ETFs, index funds, options, futures contracts).
4. Match the risk level (which involves both risk and opportunity) to your personal risk tolerance to identify the most practical choice.
5. Avoid the mistake of assigning the label "high risk" to a sector based on knowledge of only one product in that sector.

9. There Are No Convenient or Low-cost Ways to Trade Energy Futures

The cynical response to the futures market is to avoid it. Complexity and risk discourage many, but a myth prevails: that there are no low-cost ways to participate in this market.

Options present one low-cost alternative. These typically represent a small fraction of the cost of a futures contract, but they allow their owners the same degree of control as ownership of that contract does. The option is a form of leverage, but since the futures contract also leverages against its contractual size, an option of a futures contract is a form of "leverage on leverage." A futures contract itself leverages the cost of the energy product, and the option leverages the cost of the futures contract. Thus, options are not only a low-cost alternative, but they represent an exponential form of leverage.

However, this does not mean options are free of risk. This exponential leverage increases the profit potential, but it also increases (leverages) loss potential.

> **Tip:** Options are not appropriate for everyone. But for those who understand this market, costs are low, and some strategies are very conservative. This solves much of the perceived risk issues of futures trading.

Options also get around the relatively high risk of short selling in futures. Most investors understand that going short presents a high risk, but with options you can play a bearish position while remaining on the low-cost long side of a trade. A call allows you to profit if and when the underlying futures contract rises in value, but what if you believe the underlying will fall? In that case, you can buy a put. The put increases in value when the underlying falls in value.

In other words, options are a low-cost and low-risk alternative to futures contracts. As the owner of an option, you can never lose more than the cost of the option itself, a far lower risk than shorting the underlying and facing the risk of unknown losses. If you sell options on futures, you face the same risk as with shorting futures in the sense that losses can occur rapidly. However, as long as you stay on the long side with both calls and puts, your market risk is drastically reduced, and you can take either bullish or bearish positions. You can exercise the option but as a long holder you are never obligated to do so. Most long options traders sell their contracts rather than exercising them. This may represent a profit or a small loss, but the only alternative, other than exercise, is to wait for the option to expire worthless, which represents a 100% loss of the option premium.

10. Only High-Risk Speculators Can Afford Trading Futures

Given the range of product avenues available, anyone can engage in the futures market. Aside from financial, index, and currency future contracts, energy futures are the most active, and there are many ways to proceed. Doubtless, speculators are involved, but so are conservative investors whose risk tolerance demands that risks are kept low and managed carefully. With stocks, ETFs, and index funds, diversification is simple, and you can even beat inflation and taxes if you combine high dividend yield with conservative options strategies.

Key point: Speculation does not harm the market but rather tends to add liquidity to it. Blaming speculators for rising oil prices ignores the economic issues underlying those prices.

. In futures trading, if only oil and gas providers and users were involved in trading, the level of transactions would be quite low; speculation accounts for a majority of trading activity, but this is not necessarily a negative influence. In fact, the volume of speculation narrows the range between bid and ask prices and keeps the market orderly:

> Let's consider some of the principles that explain the causes of shortages and surpluses and the role of speculators. When a harvest is too small to satisfy consumption at its normal rate, speculators come in, hoping to profit from the scarcity by buying. Their purchases raise the price, thereby checking consumption so that the smaller supply will last longer. Producers encouraged by the high price further lessen the shortage by growing or importing to reduce the shortage. On the other side, when the price is higher than the speculators think the facts warrant, they sell. This reduces prices, encouraging consumption and exports, and helps to reduce the surplus.[19]

Everyone plays a role. Both long-term and highly conservative investors and high-risk, short-term traders and speculators add value to the market. The US futures market experiences a high volume of speculation buying and selling, but the market is not owned, controlled, or manipulated by speculators. Most investors know that speculators lose more than they gain, but speculation also creates a healthy profit from scarcity and helps reduce shortages. The market works because these varying interests coexist and work together.

CHAPTER 16

TEN IDEAS TO LAUNCH YOUR INVESTMENT PLAN

> One of the greatest pains to human nature is the pain of a new idea.
>
> Walter Bagehot, *Physics and Politics,* 1869

INVESTING SUCCESSFULLY REQUIRES A COMBINATION OF RESEARCH and comparison, selection of criteria for detailed fundamental analysis, application of technical timing signals, and adhering to the rules you set for yourself.

When investments do not turn out as you have planned, it often is the result of skipping these essential steps or of making decisions based on a "hot tip" from a friend or relative. But there are no shortcuts, and virtually all successful investors have one attribute worth remembering: They realize that they have to do their research and cannot rely on anyone else to replace that. Even a financial professional cannot substitute for your research and analysis of investment products and sectors. A professional can serve as a guide, but you are ultimately going to be responsible for how, when, and where you place your investment capital.

Here are ten specific ideas to help you launch a successful and profitable investment plan:

1. DEFINE YOUR INVESTING AND TRADING GOALS

As a starting point, what do you hope to accomplish? Of course, every investor wants to have more profits and fewer losses. But why are you investing? Are you ultraconservative and interested primarily in preserving your purchasing power? Are you a speculator who wants to take a chance at fast profits? Or are you somewhere in between these extremes?

The conservative investor faces a problem in the combined effects of inflation and taxes. When these two forces are considered, you have to earn the percentage of yield required to at least break even, and this might call for a higher level of risk than you are willing to assume. So you either have to raise your risk to break even or settle for earning less. If you earn less than your breakeven yield, your portfolio is not conservative. There are solutions, but they require closer hands-on investing and tracking. For example, you may focus on stocks that yield higher dividends than average, qualify, and compare these companies using a short list of powerful fundamental indicators, and then buy shares and write covered calls. If you can combine a 4% annual dividend with annualized covered call income of 10%, your overall return is 14%, higher than most breakeven requirements. Is this realistic? Yes, it is, but you need to manage your options trades appropriately to accomplish this double-digit return. This program—high-yielding stocks with covered calls—is very conservative assuming the fundamental qualifications are in place and you pick stocks wisely. There are several stocks in the energy sector meeting these requirements, so it is quite possible to make this yield happen without high risks.

2. Define Your Personal Risk Tolerance

For most investors, "risk tolerance" is a catchphrase without any depth. But do you really know your risk tolerance? If so, does your portfolio conform to it, or are you taking risks you really cannot afford?

Risk tolerance *should* be the level of risk in all of its forms (primarily market risk) that works for you based on your knowledge, experience, and capital limitations. Put in a negative light, your maximum risk tolerance is equal to the amount of capital you have at risk that you can afford to lose. In a more positive light, you need to consider the combined attributes of diversification, liquidity, market sector, and specific product selection to match risk tolerance to an action plan.

For example, if your primary goal is generating high current income while avoiding high volatility, you will be drawn toward

ETFs, mutual funds, or certain stocks, but you will be less likely to want to speculate in futures or options (unless these work as part of a conservative approach). If you are a speculator and swing trader, this dictates the kind of volatility you see as desirable in your portfolio positions. If you are conservative, you are more likely to focus on solid, well-capitalized companies with a track record of growth in revenue and earnings, increasing dividends, consistent P/E ratio, and steady or declining debt ratio—just as examples of the criteria conservative investors focused on fundamentals use.

3. List the Energy Sector Products That Are a Good Match

In the energy sector, you may compare and study stocks (in terms of dividend yield, fundamental track record, or technical volatility), ETFs, mutual funds, or commodity index funds (for diversification and professional management of a portfolio), or more advanced strategies such as options on stocks or futures; you may even consider direct trading in futures.

A "good match" between these various choices and your portfolio is defined by your risk tolerance as well as by your experience and knowledge. If you have never traded options, for example, you cannot just begin an options program in a stock portfolio without first understanding the risks. In fact, your broker will not let you trade without completing a special options trading application in which you disclose your level of experience. If you have limited capital, direct ownership of stock is limited, and you might find ETFs a better match. So what constitutes a good match is based on a combination of your risk tolerance, experience, and available capital.

4. Read, Read, Read: Find out All about Energy

A smart move for any investor is to gain information from many sources. This is true for investing and trading in general as well as for focusing on one industry, such as energy. Every industry is based on a few specific and unique attributes, and as a first step it makes sense to understand these attributes and how they affect where you invest. The energy sector presents a combination of a few essential attributes:

First, energy is a necessity. As is true also for the agricultural sector, everyone needs energy for home utilities, automobile operation, transportation, and management of all other sectors. This explains why there is such high demand for energy futures.

Second, in the energy sector, *perception* is more important than the specific economic facts, much more so than in other sectors where the market forces of supply and demand play a more direct role. . If people believe oil is about to run out or that it causes global warming or that most oil comes to the United States from the Middle East, these beliefs—whether true or not—have more to do with the value of an energy investment than the more mundane actual supply and demand of the underlying commodity.

Third, the energy sector is not viewed remotely but is influenced by passionate proponents and opponents on the basis of environmental concerns, alternative energy, and politics. The notion of the "evil of Big Oil" is widely believed in even though many of the largest oil companies have entered into partnerships with environmental interests to work together to ensure clean air and water. Thus, the black and white of "good guys and bad guys" is easy to believe in, but the truth is more complex.

5. Study the Current Financial Statements of Energy Companies

The fundamentals of any investment program may be limited to a short list of indicators, with trends followed back 10 years for example, but in addition to a short list of indicators, look closely at the financial statements of energy companies. Check some specifics, such as:

- Is the company diversified? For example, does an oil drilling company also operate refineries and delivery systems? Does the oil company pursue research into alternative energy or even other ideas outside the energy sector?
- Are the notes to the financial statements easy to understand? Are there any red flag disclosures you should research further?

- Is the annual statement easy to follow? What narrative sections of the annual report help to articulate the company's long-term growth plans?

6. Study the Market

Also look at the energy sector overall. All sectors go through cycles of growth and decline. Where is the energy sector at this moment? Because energy is a necessity, the cyclical changes are not as obvious as they are in some other sectors such as retail, travel, gaming, or autos, for example. These are affected by economic conditions; this is also true for the energy sector, but because energy is essential, the cyclical changes are more subtle and not as strongly reflected in levels of demand as in the price of the product.

The cyclical forces of the energy sector can be distinguished from those of other sectors as well. There are seasonal factors to consider each year. For example, heating oil prices may rise in the colder months; gasoline prices rise in the summer when travel increases, and any geopolitical events in the Middle East (even the threat of problems) tend to affect prices as well.

7. How Might Politics Affect Oil Industry Investment Values?

Investment in energy is not influenced by supply and demand alone. In most sectors, economic forces dominate the value and timing of investments; in the energy sector, politics plays at least as important a role in the value of your investments.

At the extreme ends of the political spectrum, conservative politicians want to drill in more places and reduce dependence on foreign oil, and liberals want to drill less and promote clean energy alternatives. Both of these extremes are unrealistic, but that isn't the point. The real point is that the side in power answers to a constituency (big business or environmental protection interests, for example), and this in turn influences whether pipelines are built, more drilling is approved, or subsidies are granted for exploration and research. Politics has a very direct influence on the energy sector and its activities as well as on the profitability of investments. Ironically, while it might appear that expanded energy activities are good for investment value, the opposite might be true. A higher level of scarcity tends to drive up prices, so that energy investments may be more profitable when political opposition to the industry is higher.

8. Diversify Effectively

Placing all of your capital into any one sector violates the principle of diversification. With this in mind, if you subscribe to the principle, you will want to allocate your capital among several different sectors. In this way, no sector-specific negative news or events will affect your entire portfolio.

Another view is that diversification is not always necessary. If you believe strongly that the energy sector is the best place to invest, you can make a conscious decision to *not* diversify and to invest everything in this one sector. This can be achieved through investment in equity (stocks or ETFs, for example), debt (exchange-traded notes or ETNs), energy futures (through commodity index funds or using options on futures), or a combination of these. In a sense, you diversify when you enter a single sector and buy different products such as these; by investing in a combination of stocks, notes, futures, and options, you achieve a form of diversification.

If you decide to keep your capital in energy or any other sector, you should also be aware of the risks involved. Diversification is a sensible idea, but that does not mean that you must diversify. It does mean that you need to be aware of the risks of an undiversified portfolio.

9. Diversify into Other Sectors

If you decide to diversify with energy and other sectors, you will want to figure out how to effectively reduce market risks. It is not true diversification when all of your holdings are subject to the same economic and cyclical forces. True diversification will include enough sectors whose economic cycles are different, so that no single economic risk affects your entire portfolio.

For example, holdings in energy, pharmaceuticals, retail, and telecommunications are subject to a range of economic forces; and diversification is effective by spreading capital among these sectors. Another method for picking sectors is to consider which ones are

Table 16.1 A diversified portfolio

Sector	Company	Symbol	yield
energy	ConocoPhillips	COP	4.3%
pharmaceuticals	GlaskoSmithKline	GSK	4.5
retail	Wal-Mart	WMT	2.5
telecommunications	AT&T	T	5.0
average			**4.1%**

Source: www.thestreet.com, yield as of June 12, 2013.

popular in the moment, which means that investment activity and market enthusiasm are high, compared to other sectors currently out of favor.

There are many ways to diversify, some more effective and practical than others. In developing a program to diversify, be aware of the need to reach your breakeven rate of return. For example, it could make sense to locate one company in each of four industries and combine dividend yield with covered call writing. This may require considering a combination of dividend yield and underlying risk attributes. For example, a list of four companies could include those listed in Table 16.1.

This list offers strong diversification by sector and also includes a combination of high dividend yields with strong fundamental quality. It shows one of many ways to achieve diversification without sacrificing better than average yield.

10. Pay Special Attention to Essential Industries (Energy, Agriculture, and Housing)

In picking sectors for investment, whether diversifying among different sectors or just trying to figure out where to invest, focus on three industries, especially for futures trading. These are energy, agriculture, and housing. These are the three basic necessities: energy, food, and shelter are needed by all people, so these sectors enjoy a distinction shared by no others. They represent products that are not going to lose their market.

In comparison, some industries are vulnerable due to changes in technology. Only a few decades ago, instant camera technology helped companies like Eastman Kodak and Polaroid grow because the technology was state of the art. Today, it is obsolete and both of these companies have lost most of their market share. Just over a century ago, a very strong industry was the livery stable business. The care and grooming of horses as well as carriages was big business until the automobile came along; by 1910 there were over 200 auto manufacturers in the United States and the wagon and carriage industry was mostly out of business.[1]

Changes in technology make many sectors vulnerable. The obsolescence of handheld or instant cameras, automobiles, and many other products of the past is a factor worth keeping in mind in picking sectors. However, energy, agriculture, and housing will always have high demand. The depressed housing prices of recent years do not change this. It may be true that fewer families are buying homes, but that means demand for rentals is higher and real estate investors are the beneficiaries. Just as the energy and agriculture sectors have many variations of products (stocks, futures, ETFs, commodity index funds), real estate can be invested in through stocks, futures, real estate investment trusts (REITs), ETFs, and mortgages. Demand does not go away, it just evolves over time.

* * *

Deciding how and where to invest in the energy market is a daunting task. You face so many choices, not only in the type of energy in which to invest but also in the type of product to focus on. The decision is affected by your risk profile, capital, condition of the market, and experience in investing and trading.

Energy is a potentially profitable market even though some products are quite volatile (which means they carry higher risk). With this in mind, the selection of one product or energy niche over another is a matter of personal choice, analysis, and evaluation of the risks.

Notes

1 A Brief History: Black Gold, Texas T

1. Mir Yusif Mir-Babayev, "Azerbaijan's Oil History," *Azerbaijan International*, September 2002.
2. CER News Desk, "Oil Production in US Hits Highest Level in 15 Years," *Christian Science Monitor*, September 28, 2012.
3. U.S. Department of Energy, www.doe.gov.
4. Bureau of Ocean Energy Management, Regulation, and Enforcement, www.boemre.gov.
5. F. A. Hayek, *The Constitution of Liberty* (Chicago: University of Chicago Press, 1960), pp. 369–70.
6. Steven Chu, quoted in Neil King Jr. and Stephen Power, "Times Tough for Energy Overhaul," *Wall Street Journal*, December 12, 2008.
7. Mark Whittington, "Energy Secretary Chu Admits Administration OK with High Gas Prices," *Yahoo! Contributor Network*, February 29, 2012.
8. "Our Mission," "Policy Statement," and "Brief History," www.opec.org, retrieved April 3, 2013.
9. Ibid.
10. Liam Denning, "Raging Oil Price will Burn OPEC," *Wall Street Journal*, December 9, 2012.
11. John Kemp, "OPEC Cannot Sustain High Oil Prices Forever," *Reuters*, January 23, 2013.
12. Brent Snavely, "US Announces Tough New 54.5-mpg CAFE Standard for Vehicle Fuel Economy," *Detroit Free Press*, August 28, 2012.

2 The Big Picture: Energy, the Biggest Market Sector

1. Numerous references in Herbert Abraham, *Asphalts and Allied Substances: Their Occurrence, Modes of Production, Uses in the Arts, and Methods of Testing*, 4th ed. (New York: Van Nostrand, 1938), p. 15, 43, and 1231; and in Herodotus, *The Histories, Book I*, ca. 440 BC.
2. Swedish Energy Agency, *Energy in Sweden, 2010*, "Facts and figures," table 46: "Total world energy supply, 1990–2009."

3. Statistical reports, 1990–2008, *www.iea.org.*
4. Jack Perkowski, "China Leads the World in Renewable Energy Investment," *Forbes,* July 27, 2012.
5. United States Geological Survey, *Arctic National Wildlife Refuge, 1002 Area, Petroleum Assessment 1998, Including Economic Analysis,* USGS Fact Sheet FS-028–01, April 2001.
6. United States Department of Energy, Energy Information Administration, *Analysis of Crude Oil Production in the Arctic National Wildlife Refuge,* 2008–03, 2008.
7. *World Oil Transit Chokepoints,* U.S. Energy Information Administration (www.eia.gov), December 30, 2011; and Anthony H. Cordesman, "Iran, Oil, and the Strait of Hormuz," *Center for Strategic and International Studies* (www.csis.org), March 26, 2007.
8. "Florida Freeze Blamed for Rise in Orange Juice Price," *www.AccuWeather .com,* January 5, 2011.
9. Christopher Doering, "More Drought in 2013 Threatens Midwest Farms," *Gannett Washington Bureau,* February 15, 2013.
10. "Top 10 Most Traded Commodities in the World 2013," *www.BiggOne .com,* February 8, 2013.
11. Jon Nicosia, "Bill O'Reilly Gets Heated With John Stossel Over Oil Speculation," www.mediaite.com, June 1, 2012.
12. Severin Borenstein, "Speculation is Not the Cause of High Oil Prices," University of California, Berkeley, Haas School of Business, May 6, 2012.
13. Ibid.
14. John Roach, "9,000-Year-Old Beer Re-Created From Chinese Recipe," *National Geographic News,* July 18, 2005, at http://news.nationalgeographic.com, retrieved April 10, 2013.
15. "Accelerating Industry Innovation: 2012 Ethanol Industry Outlook," Renewable Fuels Association, March 6, 2012, pp. 3, 8, and 10; and "First Commercial US Cellulosic Ethanol Biorefinery Announced," Renewable Fuels Association, November 20, 2006; and Monte Reel, "Brazil's Road to Energy Independence," *Washington Post Foreign Service,* August 20, 2006.
16. "Annual Energy Outlook 2013 with Projections to 2030: Issues in Focus (Early Release Overview)," Energy Information Association, April 2013 (http://www.eia.gov/forecasts/aeo/er/index.cfm).

3 The Many Uses of Energy Products

1. www.worldcoal.org.
2. "Natural Gas Vehicle Statistics: Summary Data 2010," International Association for Natural Gas Vehicles, December 2011 (www.iangv.org).
3. "Facts about Natural Gas Vehicles," Natural Gas Vehicles for America (www.ngv.org), retrieved April 13, 2013.
4. Ibid.

5. "How Safe are Natural Gas Vehicles?" Clean Vehicle Education Foundation, September 17, 2010.
6. Edwin Cartlidge, "Saving for a Rainy Day," *Science* 334 (November 18, 2011): 922–24. (Biomass encompasses several forms of biological matter, usually plants. It is converted into biofuels through thermal, chemical, or biochemical conversion processes.)
7. Ben Sills, "Solar May Produce Most of World's Power by 2060, IEA Says," *Bloomberg*, August 29, 2011.
8. "Renewables 2010 Global Status Report," Renewable Energy Policy Network (REN21), 2010, p. 53.
9. "Public Opinion on Global Issues," chapter 5b, Council on Foreign Relations (www.cfr.org), January 18, 2012.
10. "World Nuclear Power Reactors 2007–08 and Uranium Requirements," World Nuclear Association (world-nuclear.org), June 9, 2008.
11. "Nuclear Power in the USA," World Nuclear Association (world-nuclear.org), June 2008; and Matthew L. Wald, "Nuclear 'Renaissance' is Short on Largess," *New York Times*, December 7, 2010.
12. "Gauging the Pressure," *The Economist*, April 28, 2011.
13. Saswato R. Das, "The Tiny, Mighty Transistor," *Los Angeles Times*, December 15, 2007.

4 Opportunities and Risks

1. Oscar Wilde, *Lady Windermere's Fan*, 1892, Act 3.
2. The source of the name "Brent" is Shell UK, which named discovered fields after birds, and this field was named for the Brent Goose. A second explanation is that "Brent" is an acronym for layers of the oil fields, which are named Broom, Rannoch, Etieve, Ness, and Tarbat.
3. "Fact Sheet: US Oil Shale Resources, DOE Office of Petroleum Reserves Strategic Unconventional Fuels," U.S. Department of Energy, http://www.fossil.energy.gov/programs/reserves/npr/Oil_Shale_Resource_Fact_Sheet.pdf, retrieved April 15, 2013.
4. Tadeusz Patzek, David Pimentel, Michael Wang, Christopher Saricks, May Wu, Hosein Shapouri, and James Duffield, "The Many Problems with Ethanol from Corn: Just How Unsustainable Is It?" *The Phoenix Project Foundation* (http://www.phoenixprojectfoundation.us), retrieved April 16, 2013.
5. "Refinery Capacity Report," U.S. Energy Information Administration (www.eia.gov), August 3, 2012, 45.
6. "US Crude Oil, Natural Gas, and NG Liquids Proved Reserves," U.S. Energy Information Administration (www.eia.gov), August 1, 2012 (2010 totals).
7. Robert Johnston, "Deadliest Radiation Accidents and Other Events Causing Radiation Casualties," *Database of Radiological Incidents and Related Events*, September 23, 2007.
8. Everett F. Briggs, "Mine Disaster," *Science*, October 2, 1964, p. 14; and Davitt McAteer, *Monongah: The Tragic Story of the 1907 Monongah Mine*

Disaster, the Worst Industrial Accident in US History (Morgantown, WV: West Virginia University Press, 2007), p. 332.

9. "Historical Data on Mine Disasters in the United States," U.S. Department of Labor, at http://www.msha.gov/MSHAINFO/FactSheets/MSHAFCT8.HTM, retrieved April 16, 2013.
10. "History of Solar Energy," http://exploringgreentechnology.com, 2013.
11. Andrew Williams, "Run-of-the-River Hydropower Goes With the Flow," www.renewableenergyworld.com, January 31, 2012.
12. Tim Flannery, "Coal Can't be Clean," *Melbourne Herald Sun,* February 14, 2007.
13. Kate Galbraith, "Obama Signs Stimulus Packed With Clean Energy Provisions," *The New York Times*, February 17, 2009.
14. Siobhan Hughes, "Energy Secretary Backs Clean-Coal Investments," *The Wall Street Journal*, April 7, 2009.
15. Roger Blough (one-time CEO of U.S. Steel), in *Forbes*, August 1, 1967.

5 Trading in Energy Futures

1. "Refinery Capacity Report," U.S. Energy Information Administration (www.eia.gov), January 1, 2012.
2. "Nuclear Power Plants, Worldwide," European Nuclear Society (www.euronuclear.org), January 18, 2013.
3. Paul R. Ehrlich, quoted in *New Scientist,* December 14, 1967; *The Population Bomb* (New York: Ballantine Books, 1971), at http://www.newscientist.com/article/mg21929321.000-did-a-bet-on-metal-prices-save-the-lives-of-millions.html.
4. Futures Industry Organization (futuresindustry.org), 2011 statistics, retrieved April 22, 2013.
5. www.theice.com, retrieved April 22, 2013.
6. "Urbanization of America," www.theusaonline.com, retrieved April 22, 2013.

6 Energy Stocks

1. "Top Companies: Most Profitable," www.money.cnn.com, May 21, 2012.
2. S&P Stock Reports for 2012 results, at www.schwab.com.
3. "NYSE Daily Volume Statistics: Who Is Trading?" https://statspotting.com, May 6, 2011.
4. Ellen Brown, "Computerized Front-Running," *CounterPunch Magazine*, April 23, 2010.
5. David Easley, Marcos Lopez de Prado, and Maureen O'Hara, "The Microstructure of the Flash Crash: Liquidity, Crashes, and the Probability of Informed Trading," *Journal of Portfolio Management*, October 2010.
6. Carol Clark, "How to Keep Markets Safe in the Era of High-Speed Trading," *Essays on Issues,* Federal Reserve Bank of Chicago, October 2012.

7. Graham Bowley, "Lone $4.1 Billion Sale Led to Flash Crash in May," *The New York Times*, October 1, 2010; and Tom Lauricella, "How a Trading Algorithm West Awry," *The Wall Street Journal*, October 2, 2010.
8. Huw Jones, "Ultrafast Trading Needs Curbs: Global Regulators," *Reuters*, July 7, 2011.
9. Clark, op. cit.
10. Clark, op. cit., p. 3.
11. Alice K Ross, Will Fitzgibbon, and Nick Mathiason, "Britain Opposes MEPs Seeking Ban on High-Frequency Trading," *UK Guardian*, September 16, 2012.
12. Quarterly book reporting dividend achiever results, *Mergent's Dividend Achievers*," http://www.mergent.com/docs/press-releases/2012-dividend-achievers-press-release.pdf.

7 ETFs and Index Funds

1. "History of Mutual Funds," The Investment Funds Institute of Canada (www.ific.ca), retrieved May 1, 2013.
2. A "closed-end fund" limits the number of investors and asset size, and shares of modern closed-end funds are traded on exchanges just like stocks. The limitation tends to add to value even beyond the portfolio's asset value when the fund's management outperforms the market. The limited size is appealing especially to investors who have seen larger funds unable to compete as well as smaller ones.
3. A "load" is a sales charge, taken from the amount invested and used to compensate salespeople for generating business. In comparison, a no-load fund does not charge a sales fee (load) and relies on individuals investing on their own and without a salesperson or on referrals by financial professionals who are compensated by fees rather than by commissions.
4. A bond-based ETF is also called an ETN (exchange-traded notes) and some references to commodity-based ETFs use the designation ETC (exchange-traded commodities). However, all of these share the attributes of the broader exchange-traded fund.
5. "2013 Investment Company Fact Book," Investment Company Institute (http://www.ici.org).
6. Bob Russell, "ETFs vs. Mutual Funds: Which is Better?" *U.S. News & World Report*, January 17, 2013.
7. Karen Damato, "What Exactly Are 12b-1 Fees, Anyway?" *The Wall Street Journal*, July 6, 2010.
8. Dan Caplinger, "Mutual Funds vs. ETFs: Which Is Better?" *The Motley Fool* (www.fool.com), April 25, 2013.
9. A short ETF or ETN is set up as a bearish position, which means that the fund value increases when the value of the securities declines.
10. A leveraged ETF will change at a faster rate than the collective price changes in the basket of securities. For example, a 2X will change at twice the level of price movement, and a 3X will change at three times that level.

11. This ETF combines two attributes: It is a bearish (short) fund, which means that value grows as the components' values fall, and it is also leveraged at 3X; that is, the rate of change is three times greater than movement in the price.
12. USO Prospectus, "Overview of United States Oil Fund," January 31, 2013, at http://www.unitedstatesoilfund.com/documents/pdfs/uso-prospectus-20130429.pdf.
13. www.schwab.com, 1:50 EST listings, May 3, 2013.
14. To calculate compound interest: (a) divide the annual nominal rate by 4: 4.65 / 4 = 1.1625 (decimal equivalent is 0.011625); (b) add 1: 0.011625 + 1 = 1.011625 to arrive at the decimal equivalent of quarterly return; (c) multiply by 4 (quarters): 1.011625^4 = 1.047; (d) subtract 1: 1.047 – 1 = 0.47, or 4.7% compounded quarterly.

8 Options

1. A derivative is any security derived from the value of another "underlying" security. Its value is determined by movement in the price of the underlying. Derivatives include futures and forward contracts, swaps, and options.
2. Parantap Basu and William T. Gavin, "What Explains the Growth in Commodity Derivatives?" Federal Reserve Bank of St. Louis *Review 93* (1) (January/February 2011): 37–48; http://research.stlouisfed.org/publications/review/11/01/37–48Basu.pdf.
3. Either calls or puts can be covered by opening offsetting long positions. For example, a short put is accompanied by a long put at a different strike or at a later expiration. This provides cover but may also involve incurring a net debit.

9 Knowing Your Risk Tolerance

1. Glyn Holton, *Value-at-Risk: Theory and Practice* (Waltham MA: Academic Press, 2003, p. 299.
2. Joe Nocera, "Risk Management," *The New York Times Magazine*, January 4, 2009.
3. Kevin Dowd, *Measuring Market Risk* (New York: Wiley, 2005), p. 34.
4. Philippe Jorion, "Risk: Measuring the Risk in Value-at-Risk," *Financial Analysts Journal* 52 (1996): 47–56.
5. Dow Jones Company, at www.djindexes.com/sectors, downloaded September 6, 2013.

10 Using a Broker versus Going on Your Own

1. "Study on Investment Advisers and Broker-Dealers," Securities and Exchange Commission, January 2011, "Background," at http://www.sec.gov/news/studies/2011/913studyfinal.pdf.
2. Ibid., "Retail Investor Perceptions."

3. Robert S. Kaplan and David P. Norton, *Strategy Maps: Converting Intangible Assets into Tangible Outcomes* (Boston: Harvard Business Press, 2004), p. 10.
4. Cindy Barnes, Helen Blake, and David Pinder, *Creating and Delivering Your Value Proposition* (London: Kogan Page, 2009), p. 60.
5. Financial Industry Regulatory Authority (FINRA), Rule of Conduct 2210.
6. Certified Financial Planner Board of Standards, at www.cfp.net.
7. http://www.cfp.net/become-a-cfp-professional/professional-standards-enforcement.

11 Fundamental Analysis of the Energy Market

1. In October, 1973, OPEC put an embargo in effect in response to the support of the United States for Israel during the Yom Kippur War. This embargo led to long lines at service stations and a 400% rise in oil prices. Although the embargo lasted only one year, it led to the quest for alternative energy.
2. David Biello, "Has Petroleum Production Peaked, Ending the Era of Easy Oil?" *Scientific American*, January 25, 2012.
3. Bryan Walsh, "The IEA Says Peak Oil Is Dead. That's Bad News for Climate Policy," *Time*, May 15, 2013.
4. The standard for reporting in the United States has been GAAP for many years. However, by agreement between the Financial Accounting Standards Board (FASB) based in the United States and the International Accounting Standards Board (IASB), GAAP is scheduled to be merged with the International Financial Reporting Standards (IFRS) system by 2015. Source: A. Ali and L. Hwang, "Country-specific Factors Related to Financial Reporting and the Value Relevance of Accounting Data," *Journal of Accounting Research* 38 (2000): 1–21.
5. "Nearly Half of Corn Devoted to Fuel Production Despite Historic Drought," *Bloomberg View*, August 8, 2012.
6. Anthony H, Cordesman, "US Strategic Interests in the Middle East and the Process of Regional Change," Center for Strategic and International Studies (CSIS), August 1, 1996.
7. Aaron David Miller, "The Politically Incorrect Guide to US Interests in the Middle East," *Foreign Policy (FP)*, August 15, 2012, at www.foreignpolicy com.

12 Technical Analysis of the Energy Market

1. Standard deviation is calculated in six steps: (1) calculate the simple average of a number of sessions; (2) find the deviation, consisting of the closing price minus the average; (3) square each session's deviation; (4) add the squared deviations; (5) divide the total by the number of periods to find the average deviation; and (6) calculate the square root of the average to arrive at standard deviation.

2. Anne-Marie Baiynd, *The Trading Book: A Complete Solution to Mastering Technical Systems and Trading Psychology* (New York: McGraw-Hill, 2011), p. 272.
3. Adam Grimes, *The Art and Science of Technical Analysis* (Hoboken NJ: Wiley, 2012. pp. 196–98.
4. J. Welles Wilder, *New Concepts in Technical Trading Systems* (McLeansville, NC: Trend Research, 1978).
5. Joseph E. Granville. *Granville's New Key to Stock Market Profits* (Whitefish, MT: Literary Licensing, 1963 and 2011).
6. Wilder, op.cit.
7. Constance Brown, *Technical Analysis for the Trading Professional,* 2nd ed. (New York: McGraw-Hill, 2011).
8. Neil F. Chapman-Blench, *Traderevolution: Training for Traders* (Milton Keynes, UK: AuthorHouse, 2012), p. 153.
9. An exponential moving average (EMA) is formulated as a weighted average. Greater weight is given to the latest entries in the field, but unlike with other formulas, no extra calculation is required; each new entry is applied against the EMA formula, making it simple to calculate.
10. Thomas Bulkowski, *Encyclopedia of Candlestick Charts* (Hoboken, NJ: Wiley, 2008), chapters 35 (bearish engulfing) and 36 (bullish engulfing).

13 How Energy Trades Are Taxed?

1. Publication 550, at www.irs.gov/publications/p550.
2. "Are you Ready for the New Investment Tax?" *The Wall Street Journal,* April 27, 2013.
3. One exception to the rule banning passive losses is real estate. Investors can deduct up to $25,000 per year in losses from rental income as long as they meet some rules, including a requirement that they actively participate in managing their properties.
4. Based on the 33% federal plus 12.3% rate in California for those earning less than $1 million (1% more for those earning over $1 million).
5. Source: www.irs.gov.
6. Bill Bischoff, "Capital Gains: At What Rate Will Your Sale Be Taxed?" *SmartMoney,* February 4, 2013.
7. To be treated as a qualified foreign corporation, one of three conditions has to be met: (1) the company is incorporated in a United States territory; (2) the company operates under an income tax treaty with the United States; or (3) the company's stock is traded on a market within the United States.
8. Source: www.irs.gov, Publication 17.

14 Ten Mistakes Investors and Traders Make

1. "No Need to Panic about Global Warming: There's no Compelling Scientific Argument for Drastic Action to 'Decarbonize' the World's

Economy," *The Wall Street Journal*, September 9, 2013, at http://online.wsj.com/article/SB10001424052970204301404577171531838421366.html
2. "The Pitfalls of Diversification," *Forbes*, June 12, 2012.
3. M. Alex Johnson, "Mechanics See Ethanol Damaging Small Engines," *MSNBC*, August 1, 2008.
4. Ibid.

15 Ten Myths of the Energy Industry and Futures Market

1. Bryan Walsh, "The IEA Says Peak Oil Is Dead. That's Bad News for Climate Policy," *Time*, May 15, 2013.
2. U.S. Energy Information Administration (EIA), at www.eia.gov.
3. "How Much Oil is Produced in Alaska and Where Does it Go?" www.eia.gov, retrieved June 7, 2013.
4. Frank H. Murkowski, "An Honest Look at the Facts Surrounding ANWR Drilling," *The Seattle Times*, March 31, 2005.
5. "Oil Shale/Tar Sands Guide," Argonne National Laboratory and U.S. Department of the Interior, Bureau of Land Management, at http://ost-seis.anl.gov/guide, retrieved June 7, 2013.
6. J. T. Bartis, T. LaTourrette, L. Dixon, D. J. Peterson, and G. Cecchine, *Oil Shale Development in the United States, Prospects and Policy Issues* (Santa Monica CA: RAND Corporation, 2005), 9, at www.rand.org/content/dam/rand/pubs/monographs/2005/RAND_MG414.pdf.
7. Mary Griffiths, Report: "Troubled Waters, Troubling Trends," Pembina Institute, May 2006, p. 16, at http://www.pembinafoundation.org/pubs.
8. Jon Nicosia, "Bill O'Reilly Gets Heated With John Stossel Over Oil Speculation," *Mediaite*, June 1, 2012, at www.mediaite.com.
9. "Interim Report on Crude Oil," *Interim Report on Crude Oil CFTC* (Washington D.C.: Interagency Task Force on Commodity Markets, July 2008), p. 3, at http://www.cftc.gov/ucm/groups/public/@newsroom/documents/file/itfinterimreportoncrudeoil0708.pdf.
10. Robert J. Samuelson, "The Fallacy of Blaming Oil Speculators," *The Washington Post*, May 2, 2012.
11. See chapter 1 above, figure 1.5, in which the breakdown was reported as 33.5% petroleum, 20.9% natural gas, 26.8% coal, and 18.8% all others.
12. "BP Statistical Review of World Energy 2012," June 2012, p. 1, http://www.bp.com/content/dam/bp/pdf/Statistical-Review-2012/statistical_review_of_world_energy_2012.pdf.
13. Eric Lantz, Ryan Wiser, and Maureen Hand, "The Past and Future Cost of Wind Energy," U.S. Department of Energy, National Renewable Energy Laboratory (NREL), May 2012, at www.osti.gov/bridge.
14. Dina Cappiello, "Wind Farm Bird Deaths Stir Concerns in the US," *The Huffington Post*, May 14, 2013, at www.huffingtonpost.com.

15. Alyssa Mease, "Solar Panels Used by Some School Districts Prove to be More Costly than First Thought," *The Times of Trenton*, May 28, 2013.
16. "Environmental Issues Attracting More Attention, Investment from Oil Companies," *ENFOS,* April 20, 2013, at blog.enfos.com.
17. Ahmed El-Banbi and Geoffrey Thyne, "Oil and Gas Professionals' Perspectives on Common Public Perceptions of the Industry," Society of Petroleum Engineers, *The Way Ahead,* 23 vol. 9, no. 2, (2013): 22–24, at www.spe.org.
18. Sources: S&P World Commodity Fund: www.standardandpoors.com; Goldman Sachs: Source: www.spindices.com; Rogers International: www.rogersrawmaterials.com; Reuters/Jeffries: http://thomsonreuters.com; Dow Jones UBS: www.djindexes.com/commodity.
19. Victor Niederhoffer, "Daily Speculations," *The Wall Street Journal*, February 10, 1989.

16 Ten Ideas to Launch Your Investment Plan

1. Steven Klepper, "The Evolution of the US Automobile Industry and Detroit as its Capital," Carnegie Mellon University, November 2001.

Index